Teacher's Choice

Math Regents Review

Integrated Algebra
Geometry
Algebra 2 and Trigonometry

Henry Gu

Mathematics Teacher
John Dewey High School
Brooklyn, New York

Disclaimer: The contents of this book are the author's alone and not those of the New York City Department of Education.

Author: Henry Gu
Editor: Christopher Gu

www.hsmathreview.com

ISBN-10: 1450562841   ISBN-13: 9781450562843

"Everything should be made as simple as possible, but not simpler."

\- Albert Einstein

Preface

"Less is more." When students have only six to eight weeks to review for the Regents exam and they have to remember so many topics, what can the teacher offer to help? They won't be able to review the 800 page textbooks or even the 400 page review books. Our students need an efficient review kit that is concise, yet contains all the important mathematical concepts and their applications. This book will help students remember all the key topics and build their problem solving skills through the use of examples. This review book is geared towards helping students succeed with high scores on the Regents exams.

I have already used these review sheets with my own Regents classes and I have seen firsthand that their performance is significantly higher than the statewide average. Both teachers and students like these review sheets because they are practical.

This book contains three courses in one: Integrated Algebra 1, Geometry, and Algebra 2/Trigonometry. It also serves as a handy reference guide for math teachers and college students.

To effectively use this book, students should redo all the examples to fully understand the concepts and skills. They also need to practice with at least three real Regents exams to enhance their problem solving capabilities. Graphing calculators are used intensively in the Algebra 2 and Trigonometry course. TI-83 Plus and TI-84 were used for the examples in this book.

Acknowledgement

Thanks to my son Christopher, who carefully read and edited all the content in this book. He himself achieved a perfect score on the Regents exam. Thanks to the teachers and students at John Dewey High School who have already used these review sheets for their Regents prep and have achieved excellent scores. Thanks to my wife Ruping, and my family for their unconditional love and support. And finally, thanks to you, the reader, for choosing this book as your study guide. I wish you the best of luck!

Integrated Algebra 1 Review

Contents

Geometry Review

Contents

Algebra 2 and Trigonometry Review

Contents

PART 1. Algebra

I. SETS, NUMBERS, AND OPERATIONS

1. Set and Notation

A set is a collection of distinct elements.

Set Notation:
Finite Sets:
  e.g. { 1, 2, 3, 4, 5 } , { a, b, c }
Infinite Sets:
  e.g. { 1, 2 , 3, 4, 5 ··· } ,
  $\{\frac{1}{2}, \frac{1}{4}, \frac{1}{8}, \frac{1}{16} ··· \}$
Empty Set or Null Set:
     { } or Ø
  e.g. {0} is not an empty set.

Set Builder Notation:
  e.g. { x | 0 ≤ x ≤ 10, where x is a real number }

Interval Notation:
  e.g. (2, 5) represents { x | 2 < x < 5 }
     [2, 5] represents { x | 2 ≤ x ≤ 5 }
     (2, 5] represents { x | 2 < x ≤ 5 }
     [2, 5) represents { x | 2 ≤ x < 5 }

e.g. Which of the following notation is equivalent
   to the set {1, 2, 3, 4} ?
(1). { x | 1 < x < 4, where x is a whole number }
(2). { x | 0 < x < 4, where x is a whole number }
(3). { x | 1 < x ≤ 4, where x is a whole number }
(4). { x | 0 < x ≤ 4, where x is a whole number }
Here (1) is { 2, 3 };
   (2) is { 1, 2, 3 };
   (3) is { 2, 3, 4 };
   (4) is { 1, 2, 3, 4 }
Answer: (4)

2. Operations with Sets

Universe or Universal Set:
The set of all elements under consideration.

Subset:
A set which is a part of a larger set.
e.g. { 1, 2, 3 } is a subset of { 1, 2, 3, 4, 5 }

Complement of a set:
e.g. Universal Set U = { 1, 2, 3, 4, 5, 6, 7, 8 }
    Set A = { 1, 2, 3 }
The complement of set A is denoted by $\overline{A}$ or A'.
    $\overline{A}$ = { 4, 5, 6, 7, 8 }
The complement of set A has all elements in the
universal set except the elements in set A.

Intersection of two sets: symbol ∩
The set of all elements that belong to both sets.

Union of two sets: symbol ∪
The set of all elements in either set.

e.g. A = { 1, 2, 3, 4, 5 } and B = { 2, 4, 6, 8, 10 }

   then A ∩ B = { 2, 4 }
      A ∪ B = { 1, 2, 3, 4, 5, 6, 8, 10 }

3. Numbers

Counting Numbers or Natural Numbers:
      1, 2, 3, 4, 5, ...
Whole Numbers: 0, 1, 2, 3, 4, 5, ...
Integers: ..., -3, -2, -1, 0, 1, 2, 3, ...
Consecutive Integers: n, n+1, n+2, ...
  e.g. -3, -2, -1 ;   4, 5, 6
Consecutive Even Integers: n, n+2, n+4, ...
  e.g. -8, -6, -4 ;   0, 2, 4 (zero is an even number)
Consecutive Odd Integers: n, n+2, n+4, ...
  e.g. -5, -3, -1 ;   7, 9, 11
Perfect Squares: 4, 9, 16, 25, 36, 49, 64, 81, ...
Prime Numbers: 2, 3, 5, 7, 11, 13, 17, 19, 23, ...
Rational Numbers can be written as an integer, a quotient of
two integers, a terminating or repeating decimal:
  e.g.   2, -5, 1.25, 0.333..., 2.345345...,
    $\sqrt{4}= 2$, $\frac{4}{5}$, $\sqrt{\frac{4}{9}}=\frac{2}{3}$
Irrational Numbers are decimal numbers that neither repeat
nor terminate:
      $\sqrt{3}$, 1.41421... , π
Real Numbers: All the above numbers.
Rounding: 3.456 rounded to the nearest integer is 3, to the
rearest tenth is 3.5, and to the nearest hundredth is 3.46
Absolute Value of a Number: $|5| = 5$, $|-5| = 5$,
    $|12| - |-5| = 12 - 5 = 7$

  e.g. The integers are a subset of the rational numbers.
     The rational numbers are a subset of the real numbers.
     The unoin of the rational numbers and the irrational
     numbers is the set of real numbers.

4. Properties of Operations

Commutative: $a + b = b + a, \quad ab = ba$
$$3 + 5 = 5 + 3 , \quad 3 \cdot 5 = 5 \cdot 3$$
Associative: $a + (b + c) = (a + b) + c$
$$a \cdot (b \cdot c) = (a \cdot b) \cdot c$$
$$3 + (5 + 7) = (3 + 5) + 7, \quad 3(5 \cdot 7) = (3 \cdot 5)7$$
Distributive: $a(b + c) = ab + ac$
$$3(5 + 7) = 3 \cdot 5 + 3 \cdot 7$$
Additive Identity: 0
$$x + 0 = x, \quad 0 + x = x$$
$$5 + 0 = 5, \quad 0 + 5 = 5$$
Multiplicative Identity: 1
$$x \cdot 1 = x, \quad 1 \cdot x = x$$
$$3 \cdot 1 = 3, \quad 1 \cdot 3 = 3$$
Additive Inverse: $-x$
$$x + (-x) = 0 \quad\quad -5 + 5 = 0$$
Multiplicative Inverse: $\dfrac{1}{x}$
$$x \cdot \frac{1}{x} = 1 \quad\quad 3 \cdot \frac{1}{3} = 1$$
e.g. If $\odot$ is a binary operation defined by
$a \odot b = a^2 + b^2$, then $3 \odot 4 = 3^2 + 4^2 = 25$.

5. Ratio, Proportion, and Percentage

If two ratios are equal, they are in proportion.
$$\frac{a}{b} = \frac{c}{d} \quad \text{or} \quad a \cdot d = b \cdot c$$
In a proportion, the product of the means is equal to the product of the extremes.
e.g. The ratio of the three interior angles of a triangle is 2 : 3 : 4. Find the measure of the largest angle.
$$2x + 3x + 4x = 180$$
$$9x = 180$$
$$x = 20$$
The measure of the largest angle is
$$4x = 80$$

Percent:

$$\% = \frac{1}{100} , \quad 100\% = 1$$
$$60\% = \frac{60}{100} = 0.6, \quad 6\% = \frac{6}{100} = 0.06$$
$$0.6 = 0.6 \times 100\% = 60\%$$
$$6\% \text{ of } \$50 = 0.06 \cdot \$50 = \$3$$

Percent of Increase or Decrease:

$$\frac{|\text{New Amount - Original Amount}|}{\text{Original Amount}} \cdot 100\%$$

e.g. The gas price increased to \$2.50/gal from \$2.00/gal.
$$\text{Percent of Increase} = \frac{|2.50 - 2.00|}{2.00} \cdot 100\% = 25\%$$

Tax Problems:

Tax = Base Price • Tax Rate
e.g. The tax rate in NYC is 8.5%. How much do you pay for a merchandise tagged \$50?
$$\text{Tax} = 50 \cdot 8.5\% = 50 \cdot 0.085 = \$4.25$$
$$\text{Total Amount} = 50 + 4.25 = \$54.25$$

II. ALGEBRAIC EXPRESSIONS AND OPERATIONS

1. Operations of Polynomials

Combine Like Terms:
e.g. $3x^2 + x - 8 - x^2 + 5x + 4$
$$= (3x^2 - x^2) + (x + 5x) + (-8 + 4)$$
$$= 2x^2 + 6x - 4$$

Multiply:
e.g. $3y(2x^2 + 2y^2 - 2)$
$$= 3y \cdot 2x^2 + 3y \cdot 2y^2 + 3y(-2)$$
$$= 6x^2y + 6y^3 - 6y$$

e.g. $(a + b)(a + b)$
$$= a \cdot a + a \cdot b + b \cdot a + b \cdot b$$
$$= a^2 + 2ab + b^2$$

Divide:
$$\frac{12x^4 - 4x^2}{4x^2}$$
$$= \frac{12x^4}{4x^2} - \frac{4x^2}{4x^2}$$
$$= 3x^2 - 1$$

2. Factoring Polynomials

Greatest Common Factor (GCF):
$$3x^2 + 6x = 3x(x + 2)$$
$$2y^3 - 4y^2 + 2y = 2y(y^2 - 2y + 1)$$

The Difference of Two Squares:
$$a^2 - b^2 = (a + b)(a - b)$$
$$4y^2 - 25 = (2y + 5)(2y - 5)$$

Algebra 1 Review

Trinomial:

$$x^2 + 2x - 15 = (x + 5)(x - 3)$$

Here $5 \cdot (-3) = -15$ and $5 + (-3) = 2$

Factor Completely:

$$2x^3 - 14x^2 + 20x$$
$$= 2x(x^2 - 7x + 10)$$
$$= 2x(x - 2)(x - 5)$$

3. Rational Expressions

Denominator cannot be zero.

e.g. $\dfrac{x}{x - 2}$ when $x = 2$, $x - 2 = 0$, undefined

Simplify: $\dfrac{2x^3}{x^2 - x - 12} \cdot \dfrac{x^2 - 16}{6x}$

$= \dfrac{2x^3(x + 4)(x - 4)}{(x + 3)(x - 4) \cdot 6x}$ factor the numerator

and the denominator first;

$= \dfrac{x^2(x + 4)}{3(x + 3)}$ cancel out common factors in

the numerator and the denominator

Divide: $\dfrac{\dfrac{x - 3}{2x + 1}}{\dfrac{2x - 6}{2x}}$

$= \dfrac{x - 3}{2x + 1} \cdot \dfrac{2x}{2x - 6}$ multiply inverse

$= \dfrac{(x - 3) \cdot 2x}{(2x + 1) \cdot 2(x - 3)}$

$= \dfrac{x}{(2x + 1)}$

Combine: $\dfrac{1}{x + 1} + \dfrac{x - 1}{2}$ LCD = $2(x + 1)$

$= \dfrac{2 \cdot 1}{2(x + 1)} + \dfrac{(x - 1) \cdot (x + 1)}{2(x + 1)}$

$= \dfrac{2 + x^2 - 1}{2(x + 1)}$

$= \dfrac{x^2 + 1}{2(x + 1)}$

4. Radicals

$$\sqrt{a \cdot b} = \sqrt{a} \cdot \sqrt{b} \qquad a \geq 0, b \geq 0$$
$$\sqrt{\dfrac{a}{b}} = \dfrac{\sqrt{a}}{\sqrt{b}} \qquad a \geq 0, b > 0$$

Simplify: $\sqrt{75} = \sqrt{25 \cdot 3} = 5\sqrt{3}$
$\sqrt{300} = \sqrt{100 \cdot 3} = 10\sqrt{3}$

Combine: $5\sqrt{x} + 3\sqrt{x} = 8\sqrt{x}$
$\sqrt{18} - 4\sqrt{2} = \sqrt{9 \cdot 2} - 4\sqrt{2} = 3\sqrt{2} - 4\sqrt{2} = -\sqrt{2}$

Multiply: $2\sqrt{3} \cdot 4\sqrt{5} = 2 \cdot 4\sqrt{3 \cdot 5} = 8\sqrt{15}$
$3\sqrt{2} \cdot 7\sqrt{2} = 3 \cdot 7\sqrt{2 \cdot 2} = 21 \cdot 2 = 42$

Divide: $\dfrac{4\sqrt{6}}{2\sqrt{3}} = \dfrac{4}{2}\sqrt{\dfrac{6}{3}} = 2\sqrt{2}$

Rationalize: $\dfrac{1}{\sqrt{3}} = \dfrac{1}{\sqrt{3}} \cdot \dfrac{\sqrt{3}}{\sqrt{3}} = \dfrac{\sqrt{3}}{3}$

5. Exponents

$a^0 = 1 \ (a \neq 0) \qquad 5^0 = 1, \ (-5)^0 = 1, \ 1.2^0 = 1$

$x^{-n} = \dfrac{1}{x^n} \ (x \neq 0) \qquad 5^{-2} = \dfrac{1}{5^2} = \dfrac{1}{25}$

$x^a \cdot x^b = x^{a+b} \qquad 5^2 \cdot 5^3 = 5^{2+3} = 5^5$

$\dfrac{x^a}{x^b} = x^{a - b} \qquad \dfrac{8xy^3}{2xy} = \dfrac{8}{2} \cdot \dfrac{x}{x} \cdot \dfrac{y^3}{y} = 4y^2$

$(x^a)^b = x^{a \cdot b} \qquad (5^2)^3 = 5^{2 \cdot 3} = 5^6$

e.g. $(-5)^2 = (-5)(-5) = 25$
$-5^2 = -(5^2) = -25$

Scientific Notation:

$a \times 10^n \qquad 1 \leq a < 10$, n an integer

e.g. $23{,}000 = 2.3 \times 10^4$
$0.0043 = 4.3 \times 10^{-3}$

6. Evaluation of Algebraic Expressions and Formulas

e.g. If $x = 4$, $y = -3$ then
$x^2 - 4y = (4)^2 - 4(-3) = 16 + 12 = 28$

e.g. Solve for L in terms of P and W
$P = 2L + 2W$
$P - 2W = 2L$
$\dfrac{P - 2W}{2} = L \qquad L = \dfrac{P - 2W}{2}$

III. EQUATIONS AND INEQUALITIES

1. Linear Equations

$$4(x + 1) = 2x + 10 \quad \text{remove the ( ) first}$$
$$4x + 4 = 2x + 10 \quad \text{combine variables on one side, numbers on the other side;}$$
$$4x - 2x = 10 - 4 \quad \text{change the term's sign when across the = sign}$$
$$2x = 6$$
$$x = 3$$

e.g. Solve for x:
$$2ax - 5x = 2$$
$$(2a - 5)x = 2$$
$$x = \frac{2}{2a - 5}$$

e.g.
$$\frac{2}{3}x - 6 = \frac{1}{2}x + 4$$
$$4x - 36 = 3x + 24 \quad \text{multiply both sides by LCD = 6}$$
$$4x - 3x = 24 + 36$$
$$x = 60$$

Verbal Problems

Average Speed: $s = \dfrac{d}{t}$ d: Total distance, t: time

e.g. A car travels 300 miles in 5 hours, the average speed
$$s = \frac{300}{5} = 60 \text{ miles/hr.}$$

e.g. Tom drove 120 miles to his friend's house and the same distance back home. It took him 2 hours to drive there and 3 hours to drive back.
The average speed of driving out:
$$s_1 = \frac{120}{2} = 60 \text{ miles/hr.}$$
The average speed of driving back:
$$s_2 = \frac{120}{3} = 40 \text{ miles/hr.}$$
The average speed of the whole trip:
$$s = \frac{120 + 120}{3 + 2} = \frac{240}{5} = 48 \text{ miles/hr.}$$
$$\left(\text{Note: } s \neq \frac{s_1 + s_2}{2} = 50 \text{ miles/hr.} \right)$$

2. Linear Inequalities

Solving a linear inequality is the same as solving a linear equation except when both sides of the inequality are multiplied or divided by a negative number, the inequality sign is reversed.

e.g.
$$3x - 10 \geq 2$$
$$3x \geq 2 + 10$$
$$3x \geq 12$$
$$x \geq 4$$

e.g.
$$3 - 2x > 9$$
$$-2x > 9 - 3$$
$$\frac{-2x}{-2} < \frac{6}{-2} \quad \text{inequality sign reversed}$$
$$x < -3$$

3. Quadratic Equations

e.g.
$$x^2 - 10 = 3x$$
$$x^2 - 3x - 10 = 0 \quad \text{set one side equal to zero}$$
$$(x + 2)(x - 5) = 0 \quad \text{factor the trinomial}$$
$$x + 2 = 0 \quad \text{or} \quad x - 5 = 0$$
$$x = -2 \quad \text{or} \quad x = 5$$
solution set $\{ -2, 5\}$

e.g.
$$2x^2 = 5x$$
$$2x^2 - 5x = 0$$
$$x(2x - 5) = 0$$
$$x = 0 \quad \text{or} \quad 2x - 5 = 0$$
$$2x = 5$$
$$x = \frac{5}{2}$$
solution set $\{ 0, \frac{5}{2} \}$

e.g.
$$x^2 + 5 = 30$$
$$x^2 + 5 - 30 = 0$$
$$x^2 - 25 = 0$$
$$(x + 5)(x - 5) = 0$$
$$x + 5 = 0 \quad \text{or} \quad x - 5 = 0$$
$$x = -5 \quad \text{or} \quad x = 5$$

4. Rational Equations

Use Cross-Multiplying Method:

e.g. $\dfrac{x+2}{x-3} = \dfrac{3}{4}$

$4(x+2) = 3(x-3)$ cross-multiply

$4x + 8 = 3x - 9$

$4x - 3x = -9 - 8$

$x = -17$ check (denominator can not be zero)

Use LCD Method:

e.g. $\dfrac{2}{x-1} + \dfrac{1}{2} = \dfrac{4}{x-1}$ $LCD = 2(x-1)$

$4 + (x-1) = 8$ multiply LCD on
both sides

$x - 1 = 4$

$x = 5$ check: True

5. Linear System

Substitution Method:

$2x + y = 6$ (1)

$x = 3y + 10$ (2)

Substitute x by $3y + 10$ in Eq.(1):

$2(3y + 10) + y = 6$

$6y + 20 + y = 6$

$6y + y = 6 - 20$

$7y = -14$

$y = -2$

Use Eq.(2) $x = 3(-2) + 10 = -6 + 10 = 4$

Solution: $x = 4$, $y = -2$ or $(4, -2)$

Addition or Subtration Method:

$x + y = 7$ (1)

$3x - 2y = 1$ (2)

Multiply Eq.(1) by 2:

$2x + 2y = 14$ (3)

Add Eq.(3) and Eq.(2)

$2x + 2y = 14$

$\underline{3x - 2y = 1}$

$5x = 15$

$x = 3$

Substitute x by 3 in Eq. (1):

$3 + y = 7$

$y = 4$

Solution: $x = 3$, $y = 4$ or $(3, 4)$

e.g.

3 slices of pizza and 2 colas cost $6.00.

2 slices of pizza and 3 colas cost $5.25.

Find the price of one slice of pizza and one cola.

p: price of one pizza

c: price of one cola

$3p + 2c = 6$ (1)

$2p + 3c = 5.25$ (2)

Eq.(1) x 3

$9p + 6c = 18$ (3)

Eq.(2) x 2

$4p + 6c = 10.5$ (4)

Eq.(3) - Eq.(4)

$5p = 7.5$

$p = 1.5$

Replace p by 1.5 in Eq.(1)

$3(1.5) + 2c = 6$

$4.5 + 2c = 6$

$2c = 1.5$

$c = 0.75$

Answer: one slice of pizza is $1.50.
one cola is $0.75.

6. Quadratic-Linear System

e.g. $y = x^2 - 8$ (1)

$y + 5 = 2x$ (2)

From Eq. (2) $y = 2x - 5$ (3)

Substitute y by $2x - 5$ in Eq.(1):

$2x - 5 = x^2 - 8$

$x^2 - 2x - 3 = 0$

$(x - 3)(x + 1) = 0$

$x - 3 = 0$	$x + 1 = 0$
$x = 3$	$x = -1$
$y = 2(3) - 5 = 1$	$y = 2(-1) - 5 = -7$

Solution: $\{(3, 1) , (-1, -7)\}$

IV. RIGHT TRIANGLES AND TRIGONOMETRY

1. Pythagorean Theorem

$$a^2 + b^2 = c^2$$

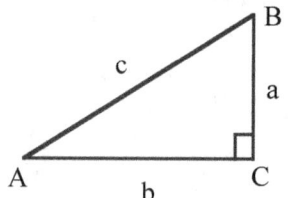

∠C is a right angle
a and b are legs
c is the hypotenuse

Pythagorean Triples:

3, 4, 5; 6, 8, 10; 9, 12, 15 etc.
5, 12, 13; 10, 24, 26 etc.

e.g. The ratio of two legs are 3:4 and the hypotenuse is 15.

Find the lengths of the two legs:
$$(3n)^2 + (4n)^2 = 15^2$$
$$9n^2 + 16n^2 = 15^2$$
$$25n^2 = 225$$
$$n^2 = 9$$
$$n = 3 \quad (n = -3 \text{ rejected})$$
$$3n = 3 \cdot 3 = 9 \quad \text{and} \quad 4n = 4 \cdot 3 = 12$$
The lengths of the two legs are 9 and 12.

2. Trigonometric Ratios

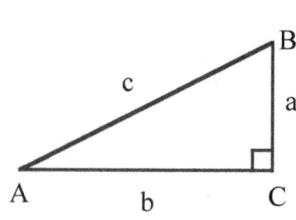

$$\sin A = \frac{\text{Opp}}{\text{Hyp}} = \frac{a}{c}$$

$$\cos A = \frac{\text{Adj}}{\text{Hyp}} = \frac{b}{c}$$

$$\tan A = \frac{\text{Opp}}{\text{Adj}} = \frac{a}{b}$$

3. Applications

Determine the right triangle and use the trigonometric ratios to solve the problem.

e.g.

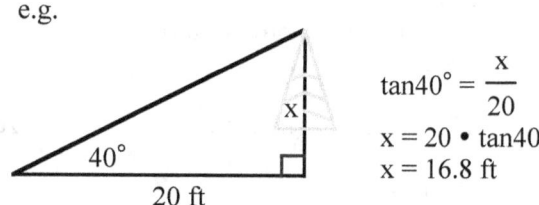

$$\tan 40° = \frac{x}{20}$$
$$x = 20 \cdot \tan 40°$$
$$x = 16.8 \text{ ft}$$

e.g.
A ladder is leaning against a vertical wall, making an angle of 60° with the ground and reaching a height of 12 feet on the wall.
Find, to the nearest foot, the length of the ladder.
Find, to the nearest foot, the distance from the base of the ladder to the wall.

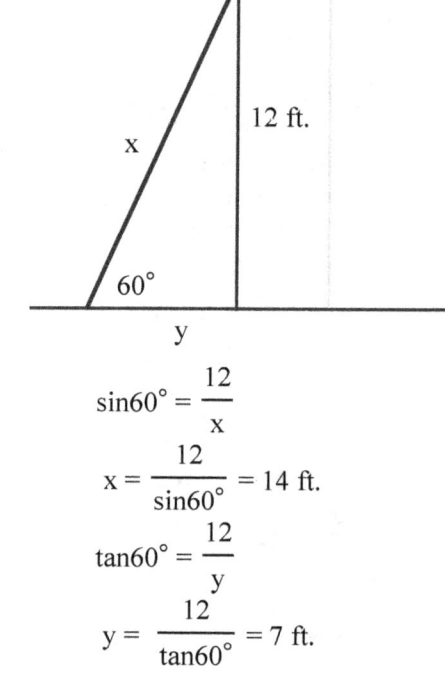

$$\sin 60° = \frac{12}{x}$$

$$x = \frac{12}{\sin 60°} = 14 \text{ ft}$$

$$\tan 60° = \frac{12}{y}$$

$$y = \frac{12}{\tan 60°} = 7 \text{ ft}$$

e.g.

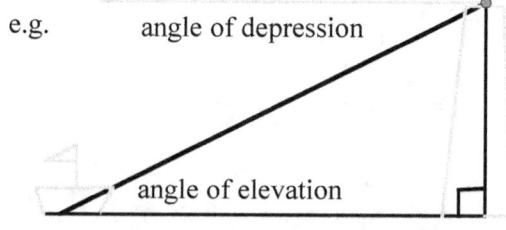

angle of depression

angle of elevation

angle of elevation = angle of depression

PART 2. Geometry and Measurements

V. BASIC GEOMETRY

1. Angles

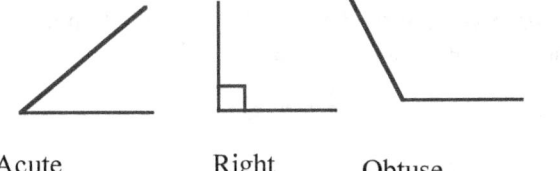

Acute   Right   Obtuse
(greater than $0°$ ($90°$)   (greater than $90°$
and less than $90°$)   and less than $180°$)

If $\angle A$ and $\angle B$ are complementary, then
     $m\angle A + m\angle B = 90$ vice versa.
If $\angle A$ and $\angle B$ are supplementary, then
     $m\angle A + m\angle B = 180$ vice versa.
A linear pair of angles are supplementary.
Vertical angles are congruent.

e.g.

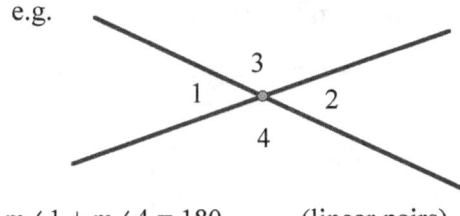

$m\angle 1 + m\angle 4 = 180$ (linear pairs)
$m\angle 2 + m\angle 4 = 180$
$\angle 1 \cong \angle 2$ and $\angle 3 \cong \angle 4$ (vertical angles)

2. Parallel Lines

Parallel lines are everywhere equidistant.

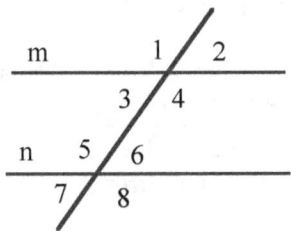

If line m $\parallel$ line n, then
$\angle 3 \cong \angle 6$, $\angle 4 \cong \angle 5$ (alternate interior angles)
$\angle 1 \cong \angle 5$, $\angle 2 \cong \angle 6$ (corresponding angles)
$\angle 3 \cong \angle 7$, $\angle 4 \cong \angle 8$
$m\angle 3 + m\angle 5 = 180$ (interior angles on the same
$m\angle 4 + m\angle 6 = 180$ side of the transversal)

3. Perpendicular Lines

Two lines are perpendicular if they form right angles.
   e.g. If $\overline{AB} \perp \overline{BC}$, then $m\angle ABC = 90$.

4. Triangles

The sum of the three interior angles $= 180°$;
exterior angle = the sum of 2 nonadjacent interior angles.

e.g.

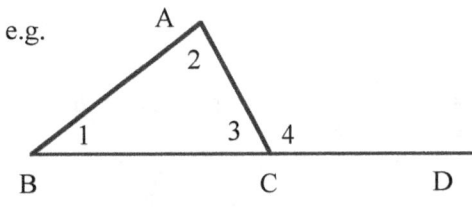

$m\angle 1 + m\angle 2 + m\angle 3 = 180$

$m\angle 1 + m\angle 2 = m\angle 4$

Any side is greater than the difference of the
other 2 sides and less than the sum of them.
$$\left| s_1 - s_2 \right| < s_3 < \left| s_1 + s_2 \right|$$

e.g. If two sides of a triangle are 3 and 5,
     then the 3rd side s_3 is
$$\left| 3 - 5 \right| < s_3 < \left| 3 + 5 \right|,$$
$$2 < s_3 < 8$$

Equilateral $\triangle$: three equal sides and three equal interior
angles ($60°$ each).
Isosceles $\triangle$: two equal sides and two equal base angles.

5. Quadrilaterals

Quadrilateral: 4-sided polygons.
Parallelogram: Opposite sides are parallel;
                Opposite sides are congruent;
                Opposite angles are congruent;
                Adjacent angles are supplementary;
                Diagonals bisect each other.
Rhombus: All the properties of the parallelogram;
            4 sides are congruent;
            Diagonals are perpendicular;
            Diagonals bisect the interior angles.
Rectangle: All the properties of the parallelogram;
            4 right angles;
            Diagonals are congruent.
Square: All the properties of the rhombus and the rectangle.

VI. GEOMETRIC MEASUREMENTS

1 yd = 3 ft , 1 ft = 12 in , 1 mile = 5280 ft
1 m = 100 cm , 1 m = 1000 mm

1. Circle

Circumference $C = 2\pi r = \pi d$ r: radius d: diameter
Area $A = \pi r^2$
e.g. When r is doubled, C is doubled and A increases 4 times.

2. Square

Perimeter $P = 4s$ s: length of the side
Area $A = s^2$

3. Rectangle

Perimeter $P = 2l + 2w$ l: length w: width
Area $A = l \cdot w$

4. Parallelogram

Perimeter P = sum of the 4 sides b: base h: height
Area $A = b \cdot h$

5. Trapezoid

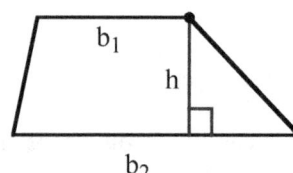

P = sum of 4 sides

$$A = \frac{b_1 + b_2}{2} \cdot h$$

6. Rhombus

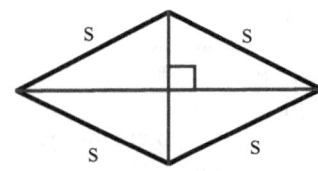

$P = 4s$

$$A = \frac{1}{2} \cdot d_1 \cdot d_2$$

d_1 and d_2 are diagonals

7. Triangle

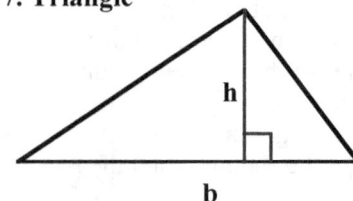

P = sum of 3 sides

$$A = \frac{1}{2} \cdot b \cdot h$$

8. Cube

Volume $V = s^3$ s : side
Surface Area $SA = 6s^2$

9. Rectangular Solid

Volume $V = l \cdot w \cdot h$ l: length, w: width, h: height
Surface Area $SA = 2lw + 2hw + 2\,lh$

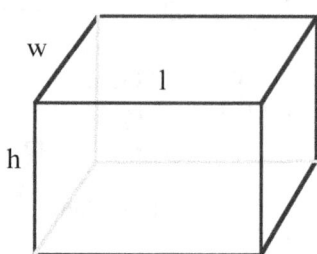

10. Circular Cylinder

Volume $V = B \cdot h$ B: area of the circular base πr^2,
h: height
Surface Area $SA = 2\pi r^2 + 2\pi rh$

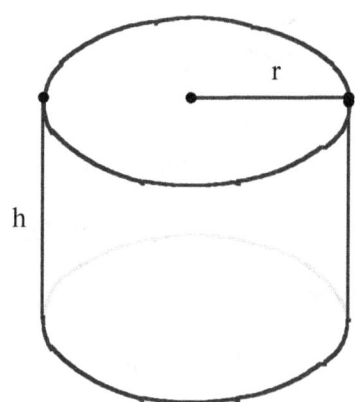

11. Error in Measurement

Absolute Error = |Measured Value - Actual Value|

$$\text{Relative Error} = \frac{\text{Absolute Error}}{\text{Actual Value}}$$

Percent of Error = Relative Error x 100%

e.g. Actual value of the side of a cube is 10.0 cm.
Measured value is 10.5 cm.
Find the relative error and percent of error in
the surface area.

$$\text{Relative Error} = \frac{\left|6(10.5)^2 - 6(10)^2\right|}{6(10)^2} = 0.1025$$

Percent of Error = 0.1025 x 100% = 10.25%

PART 3. Coordinate Geometry and Functions

VII. COORDINATE GEOMETRY AND FUNCTIONS

1. Slope

Coordinate Plane has four Quadrants I, II, III, and IV.

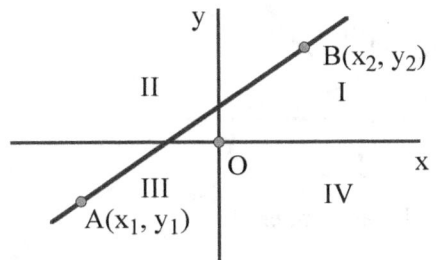

Slope $m = \dfrac{y_2 - y_1}{x_2 - x_1}$

2. Linear Function (First Degree)

A straight line can be represented as a linear function;
The graph of a linear function is a straight line.
Slope-intercept form: $y = mx + b$
where m is the slope and b is the y - intercept.

e.g.

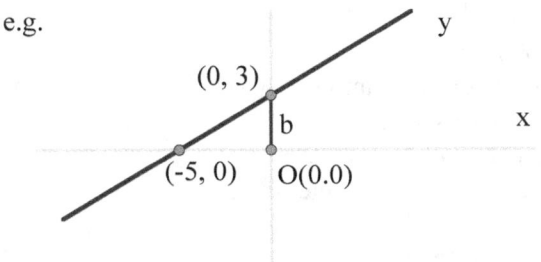

$b = 3, \quad m = \dfrac{3 - 0}{0 - (-5)} = \dfrac{3}{5}$

$y = \dfrac{3}{5} x + 3$

Two parallel lines have the same slope ($m_1 = m_2$);
The slope of a horizontal line is zero ($m = 0$);
The equation of a horizontal line: $y = b$
The slope of a vertical line is undefined;
The equation of a vertical line: $x = a$

e.g.   Find the slope and y-intercept of $3x - 2y = 12$.
        Write the equation in the slope-intercept form:

$y = \dfrac{3}{2}x - 6$,   slope $m = \dfrac{3}{2}$ and y-intercept $b = -6$

e.g. Write the equation of a line passing (3, -2) and (6, 4).

First find the slope m $= \dfrac{4 - (-2)}{6 - 3} = \dfrac{6}{3} = 2$

$y = 2x + b$,   replace x by 6 and y by 4
$4 = 2 \cdot 6 + b$ solve for $b = -8$
We have the equation of the line $y = 2x - 8$

3. Direct Variation (Ratio and Proportion)

A straight line passing through the Origin is called
Direct Variation:

$$y = mx \qquad \text{or} \qquad \frac{y}{x} = m$$

The ratio of y to x is called the constant of variation,
which is the slope of the line.

To solve a problem, use $\dfrac{x_1}{x_2} = \dfrac{y_1}{y_2}$

e.g.

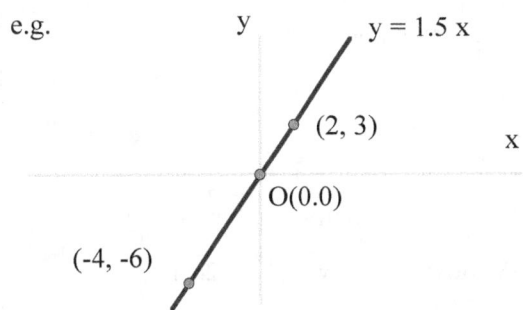

4. Absolute Value Functions

$y = |x|$

when $x < 0$
$y = -x$

when $x \geq 0$,
$y = x$

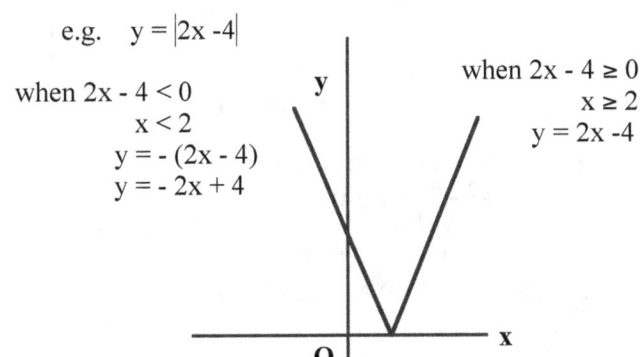

e.g. $y = |2x - 4|$

when $2x - 4 < 0$
    $x < 2$
$y = -(2x - 4)$
$y = -2x + 4$

when $2x - 4 \geq 0$
    $x \geq 2$
$y = 2x - 4$

5. Quadratic Functions and Parabolas

General Form:

$$y = f(x) = ax^2 + bx + c \qquad \text{where } a \neq 0$$

(1). Axis of Symmetry: $x = -\dfrac{b}{2a}$

(2). Vertex (Turning Point): (x, y)

$$x = -\dfrac{b}{2a}$$

use this value of x to compute $y = ax^2 + bx + c$

(3). Opening:

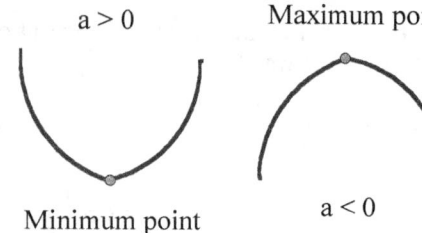

a > 0

Maximum point

Minimum point

a < 0

e.g. $y = x^2 - 2x - 3$
$a = 1, \ b = -2, \ c = -3$

(1). Axis of Symmetry: $x = \dfrac{-b}{2a} = \dfrac{-(-2)}{2(1)} = 1$

(2). Vertex: $x = 1, \ y = (1)^2 - 2(1) - 3 = -4$
Vertex: $(1, -4)$

(3). Opening: $a = 1 > 0$, upward
It has a minimum of -4 at $x = 1$

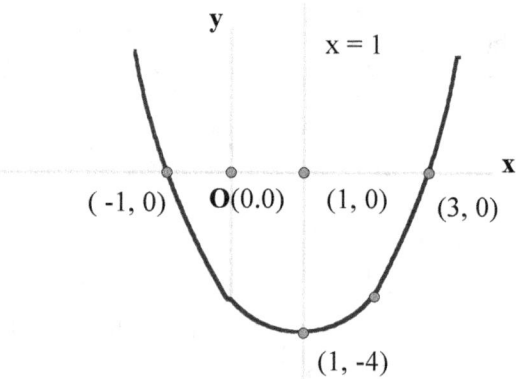

y

x = 1

x

$(-1, 0)$ O(0.0) $(1, 0)$ $(3, 0)$

$(1, -4)$

(4). Find the roots of the equation:
$$x^2 - 2x - 3 = 0$$
$$(x - 3)(x + 1) = 0$$
$$x = -1, \quad x = 3$$
x-intercepts $(-1, 0)$ and $(3, 0)$

6. Exponential Functions

$$y = a^x \qquad \text{where } a > 0 \text{ and } a \neq 1$$

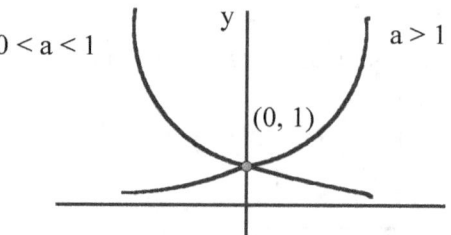

$0 < a < 1$

y

a > 1

$(0, 1)$

(1). Domain: $\{ x \mid x : \text{all real numbers} \}$
Range: $\{ y \mid y > 0 \}$
(2). $a > 1$, the function is increasing;
$a < 1$, the function is decreasing.
(3). when $x = 0, \ y = 1$
the graphs of the exponential functions passing the point $(0, 1)$
(4). x-axis is the horizontal asymptote.

General Form:

$$y = k \bullet a^x \qquad \text{where } a > 0 \text{ and } a \neq 1$$
$$k \text{ is a constant}$$

Exponential Models:

e.g. $A = A_0 2^{\frac{t}{96}}$

A_0 is the original amount. (when $t = 0$)

A is the amount at time t.

If the original amount $A_0 = 250$,
find the amount A when the time is 24.

$$A = 250 \bullet 2^{\frac{24}{96}}$$
$$A = 250 \bullet 2^{0.25} = 297.3$$

e.g. A new car will depreciate at a rate of 8% per year. If a new car is worth $15,000, how much will it be worth after 3 years?

$$A = A_0(1 - 8\%)^t$$
$$= 15,000(0.92)^3$$
$$= 11,680$$

Answer: The car will be worth $11,680 after 3 years.

7. Graphic Solutions of System of Equations

Linear System:

e.g. $x + y = 7$
         $2x - y = 2$
rewrite in the slope-intercept form
         $y = -x + 7$
         $y = 2x - 2$

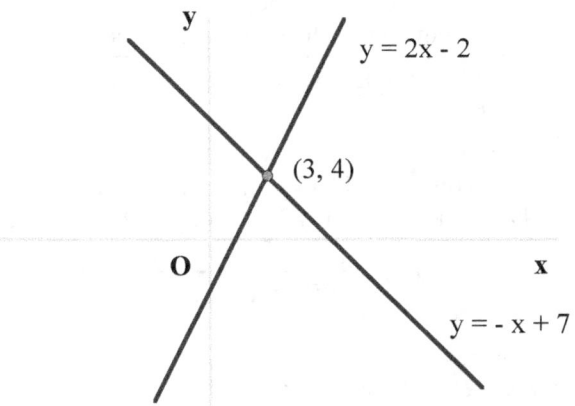

The solution of the system of equations is the intersection point (3, 4) of the two lines.

Quadratic-Linear System

e.g. $y = x^2 - 8$ (1)
         $y + 5 = 2x$ (2)
rewrite Eq. (2) in the slope-intercept form
         $y = 2x - 5$

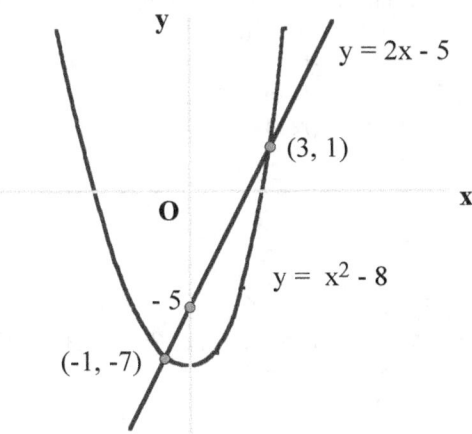

The solution of the system of equations are the intersection points (3, 1) and (-1, -7)

8. Linear Inequalities

e.g. The solution of $y < x + 2$ is the region under $y = x + 2$;

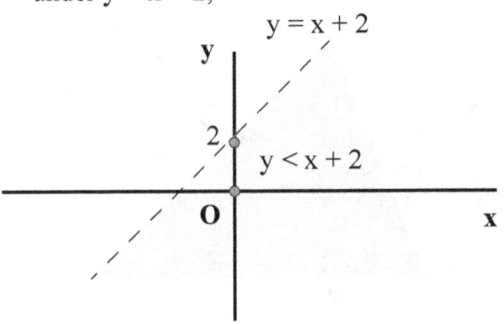

e.g. The solution of $y \geq x + 2$ is the region above $y = x + 2$ and including the line $y = x + 2$

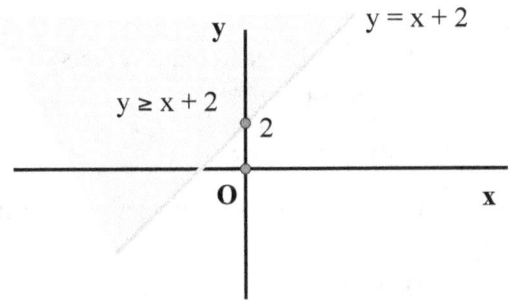

The System of Linear Inequalities

e.g. The solution of the system
         $x > 2$ (1)
         $2x - y \geq 6$ (2)
rewrite (2) in the slope-intercept form
         $-y \geq -2x + 6$
         $y \leq 2x - 6$ inequality sign reversed

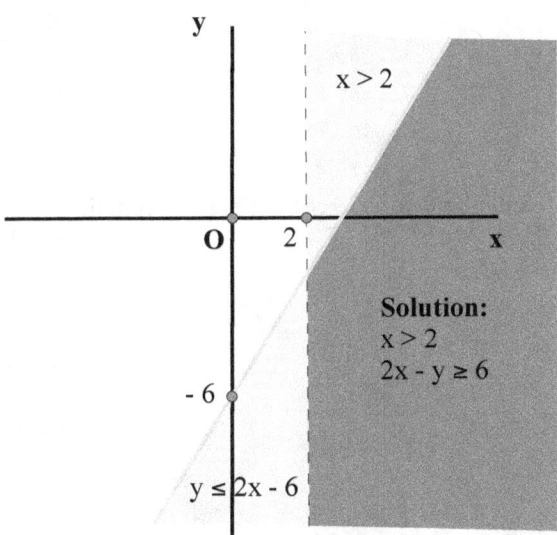

Solution:
$x > 2$
$2x - y \geq 6$

$y \leq 2x - 6$

9. Relations and Functions

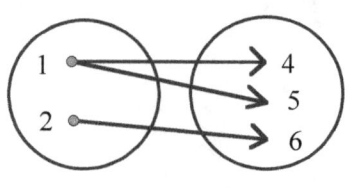

Relation (Not a Function)   Function

A function is a special relation.
The first element in the ordered pairs can not repeat in a function.
e.g. $\{(2, 1), (3, 1), (4, 3), (5, 4)\}$ is a function.
  $\{(1, 2), (1, 3), (3, 4), (4, 5)\}$ is not a function.

The **Function Notation** of $y = x^2 + 1$ is $f(x) = x^2 + 1$
  $f(2) = (2)^2 + 1 = 5$

Domain and Range:
e.g. $y = x^2$ Domain: $\{x \mid x \text{ all real numbers}\}$
              Range: $\{y \mid y \geq 0\}$
e.g. $y = \sqrt{x}$ Domain: $\{x \mid x \geq 0\}$
              Range: $\{y \mid y \geq 0\}$
e.g. $y = \dfrac{1}{x^2 - 9}$ Domain: $\{x \mid x \text{ all real numbers except } \pm 3\}$
e.g. $y = \dfrac{1}{\sqrt{x - 3}}$ Domain: $\{x \mid x > 3\}$

Vertical Line Test:
If any vertical line intersects the graph at only one point, then the relation is a function.
e.g.   Graph $y = x^2$ is a function

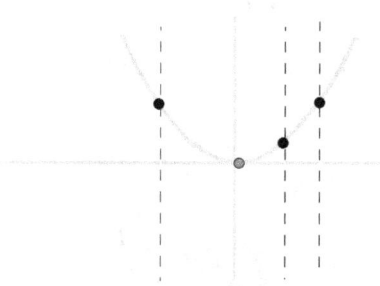

$y^2 = x$ is equivalent to $y = \pm\sqrt{x}$. It is not a function.

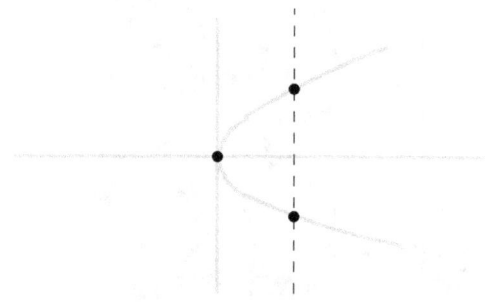

VIII. PROBABILITY

1. Probability and Counting Principle

Sample Space: all possible outcomes
Event: the favorable outcomes

(1) The probability of a simple event
$$P(E) = \frac{\text{number of the outcomes of the event}}{\text{nubmer of the outcomes of the sample space}}$$
$$P(E) = \frac{n(E)}{n(S)}$$
e.g. A bag contains 6 black balls and 4 white balls.
  What is the probability of selecting a black ball?
$$P(\text{Black}) = \frac{n(\text{Black})}{n(\text{Sample Space})} = \frac{6}{10}$$

Impossible Case: $P(E) = 0$
Certain Case: $P(E) = 1$
Negation: $P(\text{Not } E) = 1 - P(E)$

e.g.   If $P(\text{rain}) = 30\%$,
    then $P(\text{Not rain}) = 1 - P(\text{rain}) = 70\%$

(2) The probability of a single event with two conditions:
    $P(A \text{ and } B)$ meets both conditions
e.g. 52 cards , $P(K \text{ and red}) = \dfrac{2}{52}$

(3) The probability of a single event that satisfies
    condition A or condition B
    $P(A \text{ or } B) = P(A) + P(B) - P(A \text{ and } B)$
e.g.   52 cards , $P(K \text{ or red})$
    $= P(K) + P(\text{red}) - P(K \text{ and red})$
    $= \dfrac{4}{52} + \dfrac{26}{52} - \dfrac{2}{52}$
    $= \dfrac{28}{52}$

For disjoint sets A and B,
we have $P(A \text{ and } B) = 0$, then $P(A \text{ or } B) = P(A) + P(B)$
  e.g.   52 cards , $P(K \text{ or } J)$
    $= P(K) + P(J)$
    $= \dfrac{4}{52} + \dfrac{4}{52}$
    $= \dfrac{8}{52}$
    here $P(K \text{ and } J) = 0$

(4) **Counting Principle** (2 or more activities)

If the first activity can occur in M ways and the second activity can occur in N ways, then both activities can occur in M•N ways.

e.g. 3 doors to a building, 2 stairways to the second floor. There are 3•2 = 6 different ways to go.

Tree Diagram:

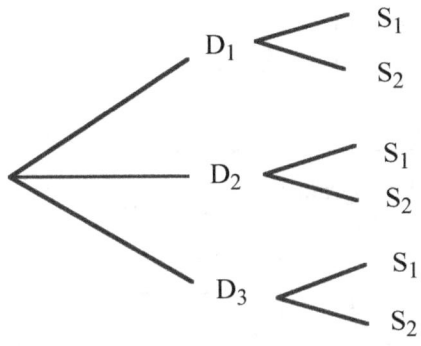

Sample Space:
$\{(D_1, S_1), (D_1, S_2), (D_2, S_1), (D_2, S_2), (D_3, S_1), (D_3, S_2)\}$

(5) Counting Principle for Probability

When A and B are indepenent events, the compound event of A and B has the probability
$$P(A, B) = P(A) \bullet P(B)$$

e.g. 4 students. The probability of the tallest one in the first place (A) and the shortest one in the last place (B)
$$P(A, B) = P(A) \bullet P(B) = \frac{1}{4} \bullet \frac{1}{3} = \frac{1}{12}$$

2. Permutation and Combination

In a **permutation** the order of the objects is important.
(1) The permutaion of n objects taken n at a time
$$_nP_n = n! = n(n-1)(n-2) \ldots 2\bullet1$$

e.g. Five letters A, B, C, D, E have 5! different arrangements. (5! = 5•4•3•2•1 = 120)

(2) The permutation of n objects taken n at a time with r items identical
$$\frac{n!}{r!}$$
e.g. Five letters COLOR have $\frac{5!}{2!}$ different arrangements.
$$(\frac{5!}{2!} = \frac{120}{2\bullet1} = 60)$$

(3) The permutation of n objects taken r (r < n) at a time
$$_nP_r = n(n-1)(n-2)\ldots \qquad (\text{r factors})$$

e.g. How many different arrangements of 1st, 2nd, and 3rd place are possible for 10 students?
$$_{10}P_3 = 10\bullet9\bullet8 = 720 \qquad (\text{3 factors})$$

In a **combination** the order of the objects does not matter.

e.g. (A, B, C) and (C, B, A) are considered same.

(4) The combination of n objects taken r at a time
$$_nC_r = \frac{_nP_r}{r!} \qquad (r \le n)$$
$$_nC_n = 1, \quad _nC_0 = 1, \quad _nC_1 = n, \quad _nC_r = _nC_{n-r}$$

e.g. How many 3 player teams can be formed from 10 students?
$$_{10}C_3 = \frac{_{10}P_3}{3!} = \frac{10\bullet9\bullet8}{3\bullet2\bullet1} = 120$$

e.g. $_{50}C_{48} = _{50}C_2 \qquad$ (to simplify the calculation)

IX. STATISTICS

1. Statistics (Univariate Data)

Analyze Data:
First arrange the data in numerical order.

$$\text{Mean} = \text{Average} = \frac{\text{sum of the data values}}{\text{number of the data items}}$$

Median: the middle value when the data arranged in order
Mode: the value that appears most often
Range: the difference between the highest value and the lowest value
Percentile: a number that tells what percent of the total number of the data values are less than or equal to a given data point
1st Quartile (25th percentile): the middle value of the lower half set of the data, aka. **Lower Quartile**
2nd Quartile (50th percentile): the median, aka. **Middle Quartile**
3rd Quartile (75th percentile): the middle value of the upper half set of the data, aka. **Upper Quartile**

Outliers: Some data points far outside most of the points in the data set.
Outliers can strongly affect the mean value. When outliers exist, use median to represent the central tendency of the data.

Circle Graph:
The whole circle (360°) represents 100% of the data.
The measure of the central angle is equal to the percentage of the 360°.
e.g. 25% of the data represented as 25% x 360° = 90°

e.g. Analyze the grades:
78, 85, 81, 95, 61, 85, 75, 88, 72, 100
First rearrange the data in numerical order:
61, 72, 75, 78, 81, 85, 85, 88, 95, 100
(make sure the number of items are same)

$$\text{Mean} = \frac{820}{10} = 82$$

$$\text{Median} = \frac{81 + 85}{2} = 83$$

(if the set has an even number of data values, take the average of the two middle values)
Mode = 85
Range = 100 - 61 = 39
Middle Quartile = Median = 83
Lower Quartile = 75
Upper Quartile = 88

Box-and-Whisker Plot :

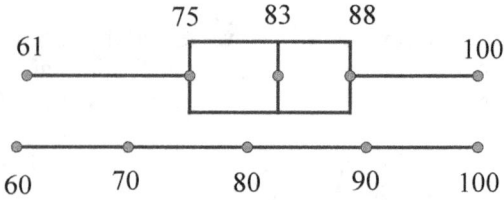

Frequency Table :

Interval	Frequency
61 - 70	1
71 - 80	3
81- 90	4
91 - 100	2

Frequency Histogram :

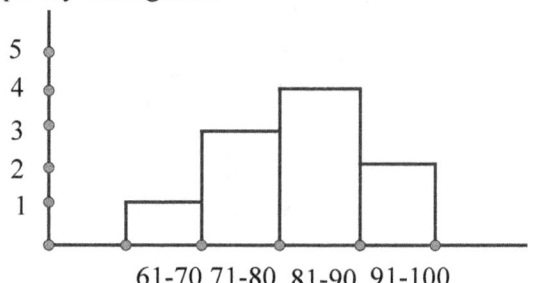

Cumulative Frequency Table :

Interval	Cumulative Frequency
61 - 70	1
61 - 80	4
61- 90	8
61 - 100	10

Cumulative Frequency Histogram :

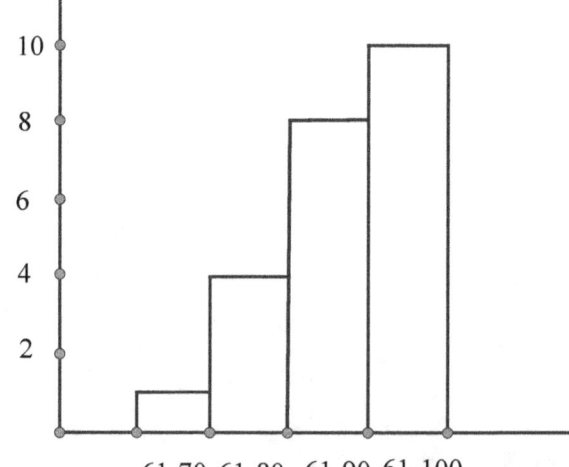

2. Statistics (Bivariate Data)

Correlation: The relationship between two sets of data

Causation: The relationship in which one variable produces an effect on the other

Regression Modeling
Linear Regression: $y = ax + b$

Correlation Coefficient r :

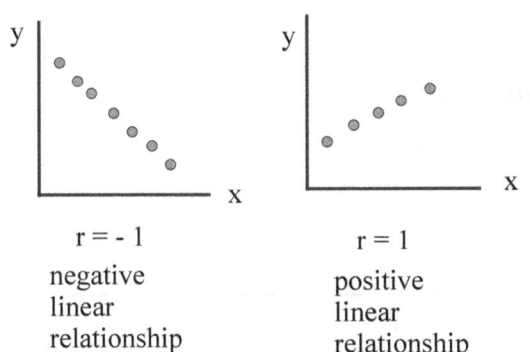

$r = -1$
negative
linear
relationship

$r = 1$
positive
linear
relationship

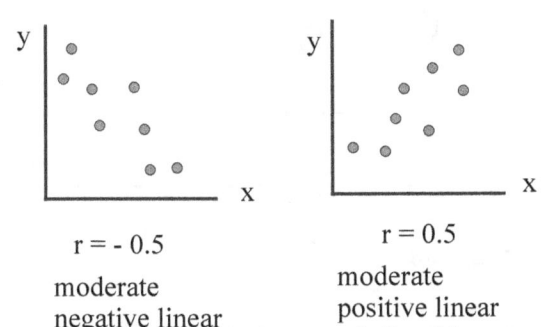

$r = -0.5$
moderate
negative linear
relationship

$r = 0.5$
moderate
positive linear
relationship

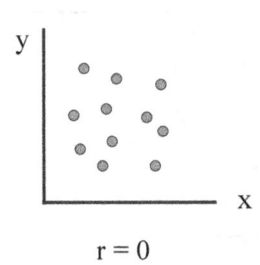

$r = 0$

no linear
relationship

Line of Best Fit (The Linear Regression)

(1). passing through the mean point $(\bar{x}, \bar{y})$
(2). the difference between the model values and the real values is the least

Use graphing calculator to find the equation of the Line of Best Fit : $y = ax + b$

e.g.

x_i	2	4	6	8	10
y_i	13	15	16	17	20

(1) Clear List L_1 and List L_2
[STAT] EDIT / 4: ClrList [ENTER] [2nd] [L_1] [,]
[2nd] [L_2] [ENTER]
(2) Enter data to L_1 and L_2
[STAT] EDIT / 1: Edit ... [ENTER]
Enter data x_i into List L_1 ;   Enter data y_i into List L_2 .
(3) Scatter Plot: [2nd] [STAT PLOT] 1: PLOT 1 [ENTER]

ON
Type:

[ZOOM] [9]

(4) Find the equation of the Line of Best Fit
and the Correlation Coefficient r :
[2nd] [CATALOG] Diagnostic On [ENTER]
[STAT] CALC / 4: LinReg(ax + b) [ENTER]
[2nd] [L_1] [,] [L_2] [ENTER]
LinReg $y = ax + b$
    $a = 0.8$ $b = 11.4$ $r = 0.98$

(5) Draw the Line of Best Fit:
[Y =] [VARS] 5: Statistics ... [ENTER] EQ /
1: RegEQ [ENTER] [ZOOM] [9]

(6) Predict the results by using the model:
Find the value of y when x = 10.5
[2nd] [CALC] 1: Values [ENTER]
X = 10.5 [ENTER]   Y = 19.8
(Adjust Window Dimensions for Extrapolation)

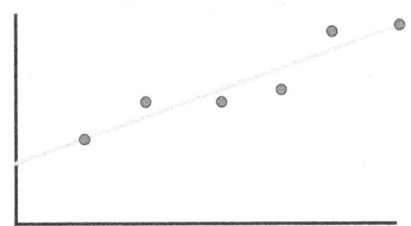

Reference Sheet

Trigonometric Ratios	$\sin A = \dfrac{opposite}{hypotenuse}$
	$\cos A = \dfrac{adjacent}{hypotenuse}$
	$\tan A = \dfrac{opposite}{adjacent}$

Area	trapezoid $A = \frac{1}{2}h(b_1 + b_2)$

Volume	cylinder $V = \pi r^2 h$

Surface Area	rectangular prism $\quad SA = 2lw + 2hw + 2lh$
	cylinder $\quad SA = 2\pi r^2 + 2\pi rh$

Coordinate Geometry	$m = \dfrac{\Delta y}{\Delta x} = \dfrac{y_2 - y_1}{x_2 - x_1}$

I. LOGIC

1. Negation: not, Symbol ~

e.g.   Statement p: I am a student.   T
        Negation ~ p: I am not a student.   F

Truth Values:

p	~p
T	F
F	T

2. Conjunction: and, Symbol ^

The conjunction p ^ q is true only when both parts are true.

Truth Value:

p	q	p ^ q
T	**T**	**T**
T	F	F
F	T	F
F	F	F

3. Disjunction: or, Symbol v

The disjunction p v q is false only when both parts are false.

Truth Value:

p	q	p v q
T	T	T
T	F	T
F	T	T
F	**F**	**F**

e.g. p: 10 is divisible by 2.   T
      q: 10 is divisible by 3.   F
      p ^ q: 10 is divisible by 2 and 10 is divisible by 3.   F
      p v q: 10 is divisible by 2 or 10 is divisible by 3.   T
      p ^ ~q: 10 is divisible by 2 and 10 is not divisible by 3. T

4. Conditional Statements

(1) **Original**   p ----> q   (If p then q.)
e.g. If it is snowing, then the school is closed.
(2) **Inverse** ~p ---> ~q (If not p then not q.)
e.g. If it is not snowing, then the school is not closed.
(3) **Converse** q ---> p (If q then p.)
e.g. If the school is closed, then it is snowing.
(4) **Contrapositive** ~q ---> ~p (If not q then not p.)
e.g. If the school is not closed, then it is not snowing.

Statements (1) and (4) are logically equivalent.
Statements (2) and (3) are logically equivalent.

5. Biconditional Statements

p <----> q (p and q have the same truth value.)
e.g. All the definitions are biconditional statements.
e.g. Two lines are perpendicular if and only if they form right angles.

II. POSTULATES

1. Postulates of Equality

(1) $a = a$ **Reflexive Property;**

e.g.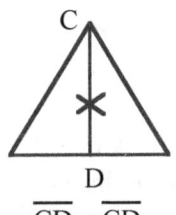

$$\overline{AB} \cong \overline{AB} \qquad \overline{CD} \cong \overline{CD}$$

(2)   If $a = b$, then $b = a$ **Symmetric Property;**
(3)   If $a = b$ and $b = c$ **Transitive Property;**
        then $a = c$
(4)   If $a = f(b)$ and $b = c$ **Substitution Postulate;**
        then $a = f(c)$
e.g.   If $a = 2b$ and $b = c$
        then $a = 2c$;
(5)   If $a = b$ and $c = d$ **Addition Postulate;**
        then $a + c = b + d$
e.g.   If $a = b$ then $a + c = b + c$;
(6)   If $a = b$ and $c = d$ **Multiplication Postulate;**
        then $ac = bd$

e.g. If $a = b$ then $\dfrac{a}{2} = \dfrac{b}{2}$; Halves of equal quantities

are equal (**Division Postulate**)

2. Postulates of Inequality

(1)   If $a > b$ and $b > c$ **Transitive Property;**
        then $a > c$

(2)   If $a > b$ and $b = c$ **Substitution Postulate;**
        then $a > c$

(3)   If $a > b$ and $c > d$ **Addition Postulate;**
        then $a + c > b + d$
e.g.   If $a > b$ then $a + c > b + c$;

(4)   If $a > b$ and $c > 0$ **Multiplication Postulate;**
        then $ac > bc$
e.g.   If $a > b$ then $2a > 2b$;

3. Partition Postulate

A whole is equal to the sum of all its parts.
e.g $AD = AB + BC + CD$
A whole is greater than any of its parts.
e.g. $AD > AB$, $AD > BD$, $AD > BC$

A   B   C   D

III. DEFINITIONS

1. Angles

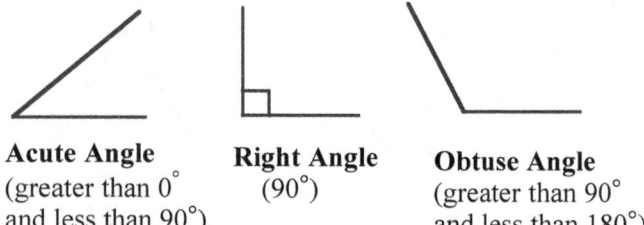

Acute Angle
(greater than $0°$
and less than $90°$)

Right Angle
($90°$)

Obtuse Angle
(greater than $90°$
and less than $180°$)

If ∠A and ∠B are **complementary**, then
$\qquad$ $m∠A + m∠B = 90$ $\qquad$ vice versa.
If ∠A and ∠B are **supplementary**, then
$\qquad$ $m∠A + m∠B = 180$ $\qquad$ vice versa.
e.g. A linear pair of angles are supplementary.

2. A **midpoint** divides a line segment into two
congruent segments.

3. A **bisector of a segment** divides the segment into
two congruent segments.

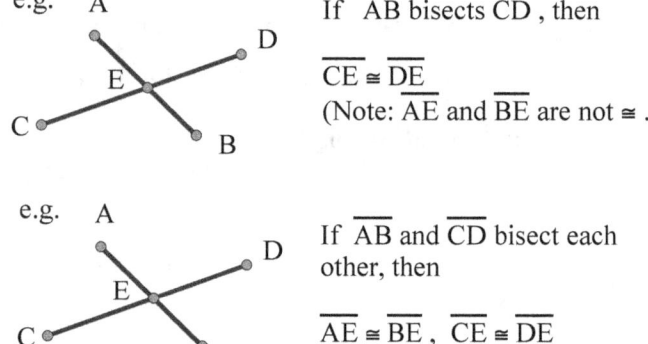

e.g. $\quad$ If $\overline{AB}$ bisects $\overline{CD}$, then

$\overline{CE} ≅ \overline{DE}$
(Note: $\overline{AE}$ and $\overline{BE}$ are not ≅.)

e.g. $\quad$ If $\overline{AB}$ and $\overline{CD}$ bisect each
other, then

$\overline{AE} ≅ \overline{BE}$, $\overline{CE} ≅ \overline{DE}$

4. A **bisector of an angle** divides the angle into two
congruent angles.

5. **Perpendicular lines** intersect to form right angles.

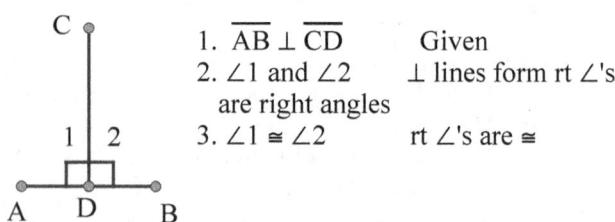

1. $\overline{AB} ⊥ \overline{CD}$ $\quad$ Given
2. ∠1 and ∠2 $\quad$ ⊥ lines form rt ∠'s
   are right angles
3. ∠1 ≅ ∠2 $\quad$ rt ∠'s are ≅

6. **Parallel lines** are in the same plane and do not
intersect.

IV. THEOREMS

1. Vertical angles are congruent.

e.g.
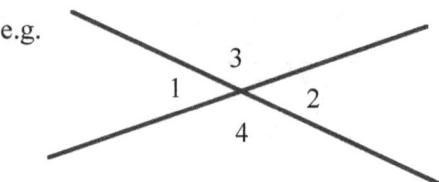

∠1 ≅ ∠2 and ∠3 ≅ ∠4 $\quad$ (vertical angles are ≅.)
$m∠1 + m∠4 = 180$ $\quad$ and $\quad m∠2 + m∠4 = 180$
(A linear pair of angles are supplementary.)

2. All right angles are congruent.

3. If two angles are congruent, then their complements
are congruent.

4. If two angles are congruent, then their supplements
are congruent.

5. Perpendicular Lines
(1) Perpendicular lines form right angles.
(2) If two lines intersect to form congruent adjacent
angles, then they are perpendicular.
(3) If a point is on the perpendicular bisector of a line
segment, then it is equidistant from the endpoints of
the line segment, and vice versa.
(4) If two points are each equidistant from the
endpoints of a line segment, these points determine the
perpendicular bisector of the segment.

6. Parallel Lines
Parallel lines are everywhere equidistant.

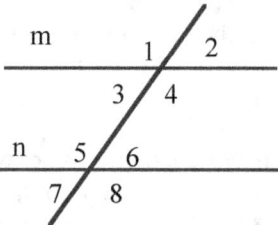

If line m ‖ line n, then alternate interior angles are ≅.
$\qquad$ ∠3 ≅ ∠6 and ∠4 ≅ ∠5
If line m ‖ line n, then corresponding angles are ≅.
$\qquad$ ∠1 ≅ ∠5, ∠2 ≅ ∠6, ∠3 ≅ ∠7, ∠4 ≅ ∠8
If line m ‖ line n, then interior angles on the same
side of the transversal are supplementary.
$\qquad$ $m∠3 + m∠5 = 180$
$\qquad$ $m∠4 + m∠6 = 180$

V. TRIANGLES

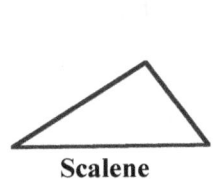  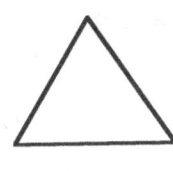

Scalene   **Isosceles**   **Equilateral**

(no congruent sides) (2 congruent sides) (3 congruent sides)

1. The sum of the three interior angles is $180°$;
The exterior angle is equal to the sum of 2 nonadjacent interior angles.

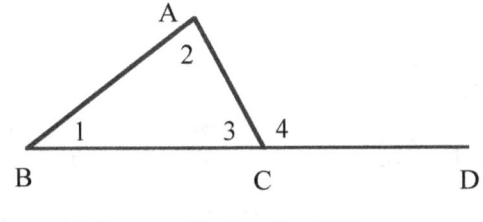

$$m\angle 1 + m\angle 2 + m\angle 3 = 180$$

$$m\angle 1 + m\angle 2 = m\angle 4$$

2. Triangle Inequalities

(1) In any triangle the greater side is opposite the greater angle, and vice versa.

(2) Any side is greater than the difference of the other 2 sides and less than the sum of them.
$$\left|s_1 - s_2\right| < s_3 < \left|s_1 + s_2\right|$$
e.g. If the two sides of a triangle are 3 and 5, then the 3rd side s_3 is $\left|3 - 5\right| < s_3 < 3 + 5$, which is $2 < s_3 < 8$

(3) Any exterior angle is greater than either nonadjacent interior angle.

3. Median, Altitude, and Angle Bisector

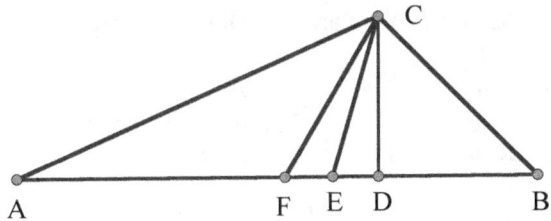

Given: $\overline{CD}$ is an altitude, $\overline{CE}$ is an angle bisector, and $\overline{CF}$ is a median.

$\overline{CD} \perp \overline{AB}$	Def. of altitude
$\angle ACE \cong \angle BCE$	Def. of angle bisector
F is the midpoint of $\overline{AB}$	Def. of median
$\overline{AF} \cong \overline{BF}$	Def of midpoint

Concurrence (Intersect in One Point)

(1) Centroid:
The medians of a triangle are concurrent.
(It is the Center of Gravity)
The centroid divides each median in the ratio 2 to 1.

(2) Incenter:
The angle bisectors of a triangle are concurrent.
(It is the Center of the Inscribed Circle
        --- equidistant from each side)

(3) Orthocenter:
The altitudes of a triangle are concurrent.
The orthocenter of an obtuse $\triangle$ is outside of the triangle.

(4) Circumcenter:
The perpendicular bisectors of the sides of a triangle are concurrent.
(It is the Center of the Circumscribed Circle
        --- equidistant from each vertex.)
The circumcenter of an obtuse $\triangle$ is outside of the triangle.

e.g.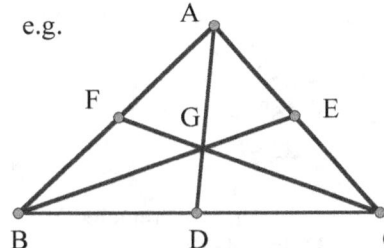

G is the centroid.
Then:
DG = x
AG = 2x
AD = 3x

4. Isosceles Triangle

Definition: A triangle that has two congruent sides.

(1) The base angles of an isosceles triangle are congruent.
(2) If two angles of a triangle are congruent, then their opposite sides are congruent.
(3) To the base of an isosceles triangle, the median, altitude, angle bisector, and perpendicular bisector coincide.

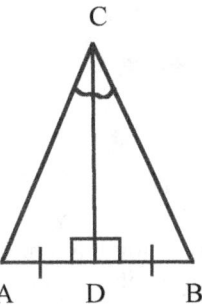

$\triangle ABC$ is isosceles with $\overline{AC} \cong \overline{BC}$,
$\overline{CD}$ is the median, altitude, angle bisector, and perpendicular bisector.

5. Right Triangle

Pythagorean Theorem

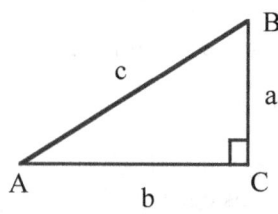

$\angle C$ is a right angle
a and b are legs
c is the hypotenuse

$$a^2 + b^2 = c^2$$

Pythagorean Triples:

3, 4, 5; 6, 8, 10; 9, 12, 15 etc.
5, 12, 13; 10, 24, 26 etc.

e.g.
The ratio of two legs are 3:4 and the hypotenuse is 15.
Find the lengths of the two legs:

$$(3n)^2 + (4n)^2 = 15^2$$
$$9n^2 + 16n^2 = 15^2$$
$$25n^2 = 225$$
$$n^2 = 9$$
$$n = 3 \quad (\, n = -3 \text{ rejected })$$
$$3n = 3 \cdot 3 = 9 \text{ and } 4n = 4 \cdot 3 = 12$$

The lengths of the two legs are 9 and 12.

Special Right Triangles

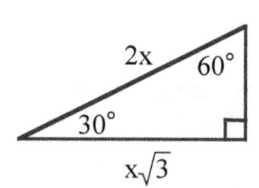 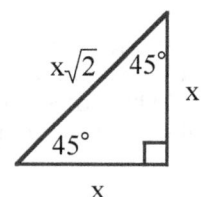

6. Congruent Triangles

SSS $\cong$
SAS $\cong$
ASA $\cong$
AAS $\cong$ (No SSA $\cong$)
HL $\cong$ (for right triangles only)
CPCTC: Corresponding Parts of Congruent Triangles are Congruent.

e.g. $\overline{AB}$ and $\overline{CD}$ bisect each other at point E.
Prove: $\triangle ACE \cong \triangle BDE$

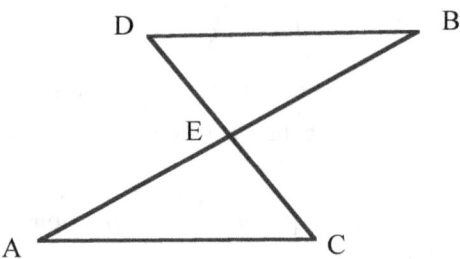

1. $\overline{AB}$ and $\overline{CD}$ bisect each other at point E	1. Given
2. $\overline{AE} \cong \overline{BE}$ and $\overline{CE} \cong \overline{DE}$	2. A bisector divides a segment into two congruent segments
3. $\angle AEC \cong \angle BED$	3. Vertical angles are congruent
4. $\triangle ACE \cong \triangle BDE$	4. SAS $\cong$

e.g. $\overline{CA} \perp \overline{AB}$, $\overline{DB} \perp \overline{AB}$, $\overline{AD} \cong \overline{BC}$
Prove: $\triangle ABC \cong \triangle BAD$

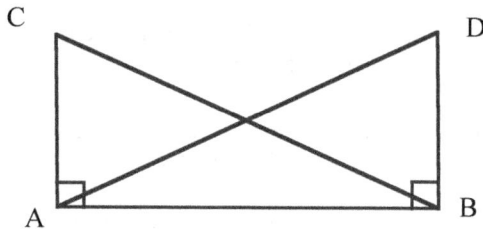

1. $\overline{CA} \perp \overline{AB}$, $\overline{DB} \perp \overline{AB}$	1. Given
2. $\angle CAB$ and $\angle DBA$ are right angles	2. $\perp$ lines form right angles
3. $\triangle ABC$ and $\triangle BAD$ are right $\triangle$s	3. Def. of the right $\triangle$
4. $\overline{AD} \cong \overline{BC}$	4. Given
5. $\overline{AB} \cong \overline{AB}$	5. Reflexive property
6. $\triangle ABC \cong \triangle BAD$	6. HL $\cong$

7. Ratios, Proportions, and Similar Triangles

(1) Ratios and Proportions

If two ratios are equal, they are in proportion.

$$\frac{a}{b} = \frac{c}{d} \quad \text{or} \quad a \bullet d = b \bullet c$$

In a proportion, the product of the means is equal to the product of the extremes.

(2) Similar Triangles
AA ~ (most often used for proof)
SAS ~
SSS ~

If two triangles are similar, then their corresponding angles are congruent and their corresponding sides are in proportion.

In two similar triangles, the ratio of the perimeters is equal to the ratio of the sides.

In two similar triangles, the ratio of the areas is equal to the square of the ratio of the sides.

e.g If $\triangle ABC \sim \triangle A'B'C'$ and $\dfrac{AB}{A'B'} = \dfrac{2}{1}$

then $\dfrac{P}{P'} = \dfrac{2}{1}$ and $\dfrac{\text{Area}}{\text{Area'}} = \dfrac{4}{1}$

Theorem

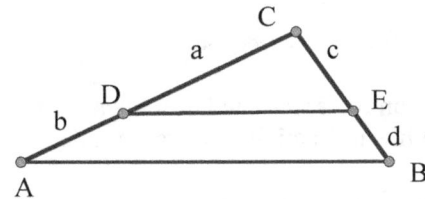

If $\overline{DE} \parallel \overline{AB}$, then $\dfrac{a}{b} = \dfrac{c}{d}$

Midsegment Theorem

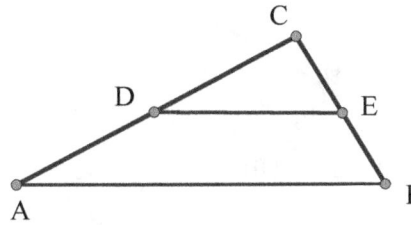

If D and E are midpoints of $\overline{AC}$ and $\overline{BC}$, then the midsegment $\overline{DE}$ is parallel to $\overline{AB}$ and is half of $\overline{AB}$.

e.g.

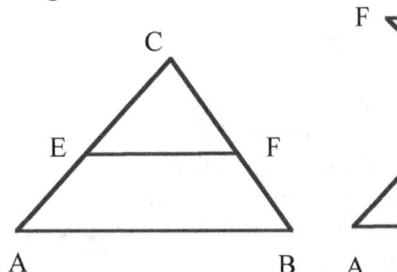

If $\overline{EF} \parallel \overline{AB}$,
then $\triangle EFC \sim \triangle ABC$ AA~

then $\dfrac{EC}{AC} = \dfrac{FC}{BC} = \dfrac{EF}{AB}$ Corresponding sides in proportion

e.g.

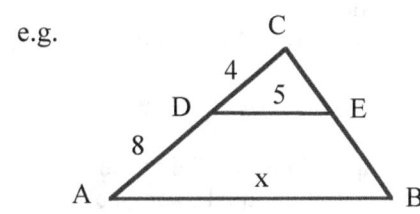

If $\overline{AB} \parallel \overline{DE}$

then $\dfrac{5}{x} = \dfrac{4}{4+8}$, $x = 15$

(Note: $\dfrac{5}{x} \ne \dfrac{4}{8}$)

(3) Proportions in the Right Triangle

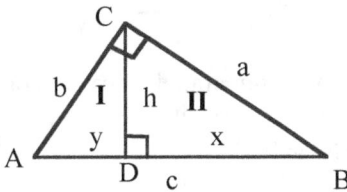

Right Triangle Altitude Theorem

(a). $\triangle \mathbf{I} \sim \triangle \mathbf{II} \sim \triangle ABC$

(b).   The altitude to the hypotenuse is the geometric mean of the two segments of the hypotenuse.
$$h^2 = x\,y \; ;$$

(c).   Each leg is the geometric mean of its projection on the hypotenuse and the whole hypotenuse.
$$a^2 = x\,(x+y) = xc \; ;$$
$$b^2 = y\,(x+y) = yc$$

VI. POLYGONS

1. Quadrilateral: a 4-sided polygon.

2. To Prove a Parallelogram:
  2 pairs of opposite sides are parallel;
  2 pairs of opposite sides are congruent;
  2 pairs of opposite angles are congruent;
  1 pair of opposite sides are parallel and congruent;
  Diagonals bisect each other.
Rhombus: All the properties of the parallelogram;
  4 sides are congruent;
  Diagonals are perpendicular;
  Diagonals bisect the interior angles.
Rectangle: All the properties of the parallelogram;
  4 right angles;
  Diagonals are congruent.
Square: All the properties of the rhombus and the rectangle.

3. Trapezoid: one and only one pair of opposite sides
  are parallel.
  The median of a trapezoid is parallel to the bases.
  The length of the median is equal to one-half the
  sum of the lengths of the bases.

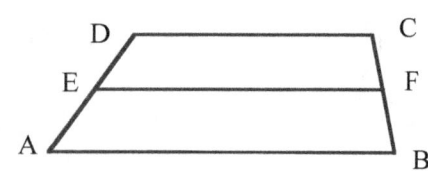

$\overline{EF}$ is the median: $\overline{EF} \parallel \overline{AB}$, $\overline{EF} \parallel \overline{CD}$
$$EF = \frac{AB + CD}{2}$$

Isosceles Trapezoid: the nonparallel sides are congruent.
  Base angles are congruent.
  Diagonals are congruent.

4. Polygons

Sum of the exterior angles = $360°$

Exterior angle of a regular polygon = $\dfrac{360°}{n}$

Sum of the interior angles = $n \cdot 180° - 360°$
$$= (n - 2) \cdot 180°$$

Interior angle of a regular polygon = $180° - \dfrac{360°}{n}$

Square (4 sides), Pentagon (5 sides), Hexagon (6 sides),
Octagon (8 sides).

VII. COORDINATE GEOMETRY

1. Slope, Midpoint, Distance, and Centroid

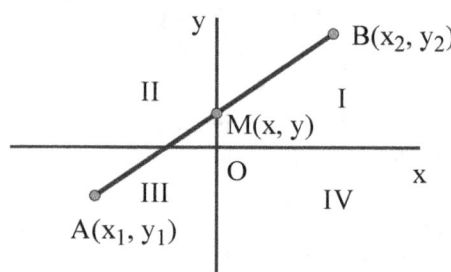

Coordinate Plane has four Quadrants I, II, III, and IV.

slope $m = \dfrac{y_2 - y_1}{x_2 - x_1}$

midpoint $M(\overline{x}, \overline{y}) = M(\dfrac{x_1 + x_2}{2} , \dfrac{y_1 + y_2}{2})$

distance $d = \sqrt{(x_2 - x_1)^2 + (y_2 - y_1)^2}$

e.g. $\overline{AB}$ has midpoint M(1,4) and one end B(3,5).
Find the coordinates of the other end A.
$$1 = \frac{x_1 + 3}{2} \quad , \quad 4 = \frac{y_1 + 5}{2}$$
  Solve for x_1 and y_1. $A(x_1, y_1) = A(-1, 3)$
e.g. The **Centroid** of a triangle is
$$(\frac{x_1 + x_2 + x_3}{3} , \frac{y_1 + y_2 + y_3}{3})$$

2. Linear Function (First Degree)

A straight line can be represented as a linear function;
The graph of a linear function is a straight line.

Slope-Intercept Form: $y = mx + b$
where m is the slope and b is the y - intercept.

Point-Slope Form: $y - y_1 = m(x - x_1)$

e.g.

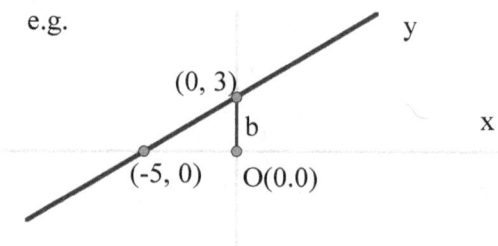

$$b = 3, \quad m = \frac{3 - 0}{0 - (-5)} = \frac{3}{5} , \quad y = \frac{3}{5}x + 3$$

Special cases:
Direct Variation: when b = 0,
the line passing through the Origin
$$y = mx$$
Vertical line: x = a
Horizontal line: y = b

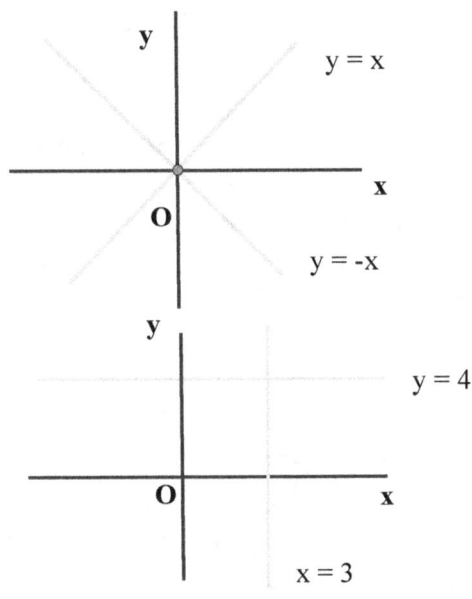

Two parallel lines have the same slope ($m_1 = m_2$);

Two perpendicular lines: $m_2 = -\dfrac{1}{m_1}$ or $m_1 \cdot m_2 = -1$;

The slope of a horizontal line is zero (m = 0);
The slope of a vertical line is undefined.

e.g. Find the slope and y-intercept of 3x - 2y = 12.
Write the equation in slope and y-intercept form:
$$y = \frac{3}{2}x - 6 \ , \ \text{slope } m = \frac{3}{2} \text{ and y-intercept } b = -6$$

e.g. Write the equation of a line through (3, -2) and (6, 4).
First find the slope $m = \dfrac{4 - (-2)}{6 - 3} = \dfrac{6}{3} = 2$
y = 2x + b ,   replace x by 6 and y by 4
$4 = 2\cdot 6 + b$ solve for b = -8
We have the equation of the line y = 2x - 8

e.g. Write the equation of a line passing through the origin
and perpendicular to the line y = 2x + 3.
Since the line passing through the origin: y = mx   (b = 0)
$$m_2 = -\frac{1}{m_1} = -\frac{1}{2} \ , \quad \text{Solution: } \ y = -\frac{1}{2}x$$

3. Coordinate Geometric Proof

(1).To prove a parallelogram:
Method 1: (Slope formula)
Two pairs of opposite sides are parallel - the same slope;
Method 2: (Midpoint formula)
Diagonals having the same midpoint bisect each other.
(2). To prove a rhombus:
(Distance formula)
4 sides have the same length;
(3). To prove a rectangle:
(Slope formula)
Opposite sides are parallel, adjacent sides are
perpendicular;
(4). To prove a trapezoid:
(Slope formula)
One pair of the opposite sides are parallel - the same slope;
and the other pair of the opposite sides are not parallel -
different slopes.

e.g. The quadrilateral ABCD has vertices A(-5, -2),
B(-5, 3), C(4, 6), and D(7, 2). Prove by coordinate
geometry that quadrilateral ABCD is an isosceles
trapezoid

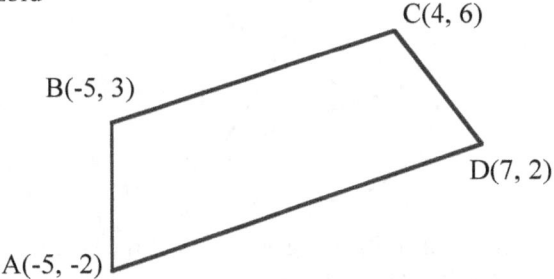

Prove:

Slope of $\overline{AD} = \dfrac{2 - (-2)}{7 - (-5)} = \dfrac{4}{12} = \dfrac{1}{3}$

Slope of $\overline{BC} = \dfrac{6 - 3}{4 - (-5)} = \dfrac{3}{9} = \dfrac{1}{3}$

Slope of $\overline{AD}$ = Slope of $\overline{BC}$ $\overline{AD} \parallel \overline{BC}$

Slope of $\overline{AB} = \dfrac{3 - (-2)}{-5 - (-5)} = \dfrac{5}{0}$ (vertical line)

Slope of $\overline{CD} = \dfrac{2 - 6}{7 - 4} = \dfrac{-4}{3}$

Slope of $\overline{AB} \neq$ Slope of $\overline{CD}$ $\overline{AB}$ is not $\parallel \overline{CD}$
Therefore ABCD is a trapezoid.

$AB = \sqrt{[-5 - (-5)]^2 + [3 - (-2)]^2} = 5$
$CD = \sqrt{(7 - 4)^2 + (2 - 6)^2} = 5$
AB = CD
Therefore ABCD is an isosceles trapezoid.

VIII. CIRCLE

1. Angles of a Circle
The degree measure of a circle is 360°.
The degree measure of a semicircle is 180°.
The degree measure of an arc is equal to the measure of the central angle that intercepts the arc.

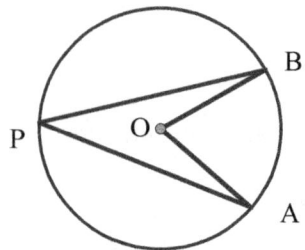

Central angle $m\angle AOB = m\overset{\frown}{AB}$

Inscribed angle $m\angle APB = \frac{1}{2}m\overset{\frown}{AB}$

Inscribed angle of a semicircle is a right angle, vice versa.

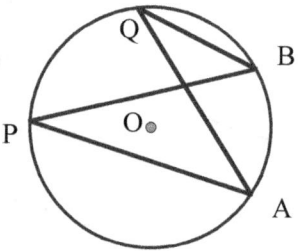

If two inscribed angles intercept the same arc, they are congruent.
$$\angle P \cong \angle Q$$

Congruent arcs have congruent chords, vice versa.

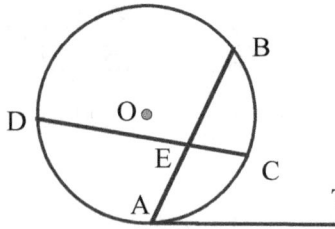

Chord - tangent angle: $m\angle BAT = \frac{1}{2}m\overset{\frown}{BCA}$

Chord - chord angle: $m\angle BEC = \frac{1}{2}(m\overset{\frown}{BC} + m\overset{\frown}{AD})$;

$$m\angle AEC = \frac{1}{2}(m\overset{\frown}{AC} + m\overset{\frown}{BD})$$

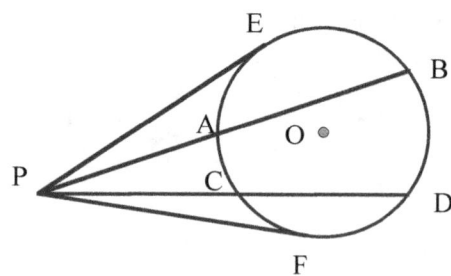

Tangent - tangent angle $m\angle EPF = \frac{1}{2}(m\overset{\frown}{EBDF} - m\overset{\frown}{EF})$

Tangent - secant angle $m\angle EPB = \frac{1}{2}(m\overset{\frown}{EB} - m\overset{\frown}{EA})$

Secant - secant angle $m\angle BPD = \frac{1}{2}(m\overset{\frown}{BD} - m\overset{\frown}{AC})$

e.g.

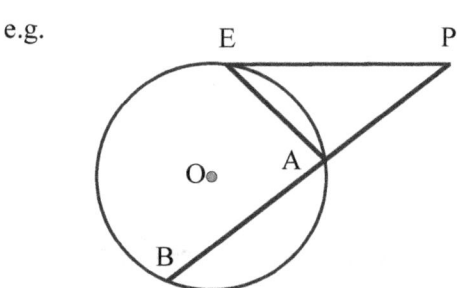

Given: $\overline{PE}$ is tangent.
$m\overset{\frown}{EA} : m\overset{\frown}{AB} : m\overset{\frown}{BE} = 2 : 3 : 4$

Find: $m\angle P$, $m\angle PEA$, $m\angle PAE$

$m\overset{\frown}{EA} : m\overset{\frown}{AB} : m\overset{\frown}{BE} = 2x : 3x : 4x$
$2x + 3x + 4x = 360 \qquad x = 40$

$m\overset{\frown}{EA} = 80$, $m\overset{\frown}{AB} = 120$, $m\overset{\frown}{BE} = 160$

$m\angle P = \frac{1}{2}(m\overset{\frown}{BE} - m\overset{\frown}{EA}) = 40$

$m\angle PEA = \frac{1}{2}m\overset{\frown}{EA} = 40$

$m\angle PAE = 180 - 40 - 40 = 100$

2. Segments of a Circle

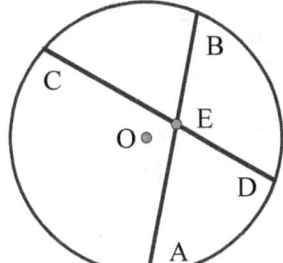

$$AE \cdot EB = CE \cdot ED$$

$$\frac{AE}{CE} = \frac{ED}{EB}$$

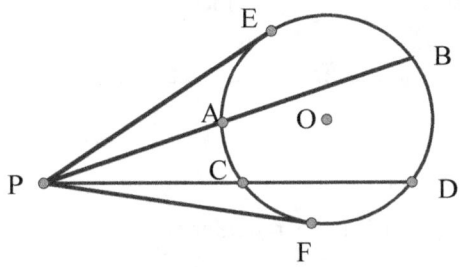

$\overline{PE}$, $\overline{PF}$ are tangent segments. $\overline{PE} \cong \overline{PF}$
$PE^2 = PA \cdot (PA + AB) = PC \cdot (PC + CD)$

$$\frac{PA}{PE} = \frac{PE}{PB} \qquad \frac{PC}{PE} = \frac{PE}{PD} \qquad \frac{PA}{PC} = \frac{PD}{PB}$$

3. Theorems

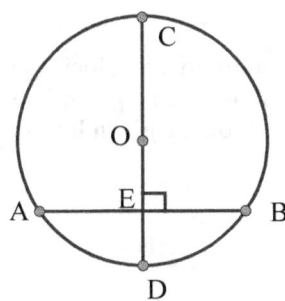

$$\overline{PA} \cong \overline{PB}$$

A tangent to a circle is perpendicular to the radius at its point of intersection.

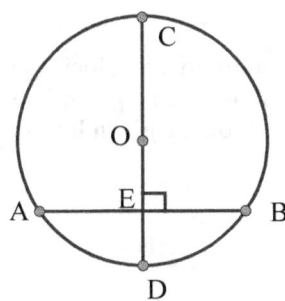

A diameter perpendicular to a chord bisects the chord and its arcs.
If $\overline{AB} \perp \overline{CD}$, then $\overline{AE} \cong \overline{BE}$ and $\overparen{AD} \cong \overparen{BD}$,
$\overparen{AC} \cong \overparen{BC}$.

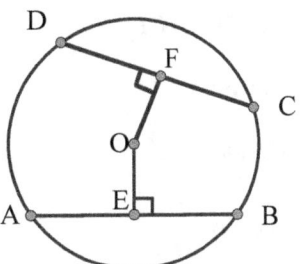

If two chords of a circle are congruent, then they are equidistant from the center of the circle, vice versa.
If $\overline{AB} \cong \overline{CD}$, then $\overline{OE} = \overline{OF}$;   or
If $OE = OF$, then $\overline{AB} \cong \overline{CD}$
If $OE < OF$, then $AB > CD$

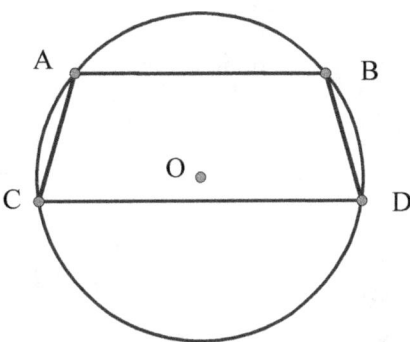

In a circle, parallel chords intercept congruent arcs between them.
If $\overline{AB} \parallel \overline{CD}$, then $\overparen{AC} \cong \overparen{BD}$

In a circle, congruent arcs have congruent chords, vice versa.
If $\overparen{AC} \cong \overparen{BD}$, then $\overline{AC} \cong \overline{BD}$

4. Common Tangents of Two Circles:

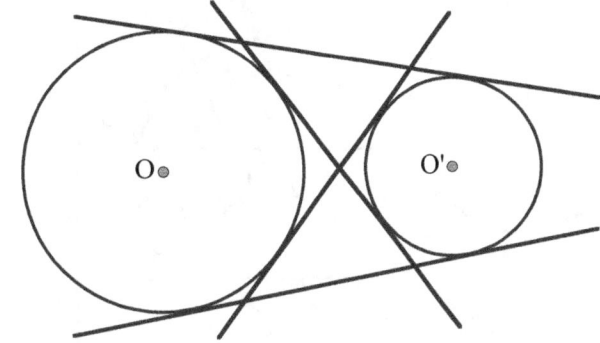

IX. CONSTRUCTIONS AND LOCI

1. Three Types of Constructions

(1) Copy a line segment or an angle.

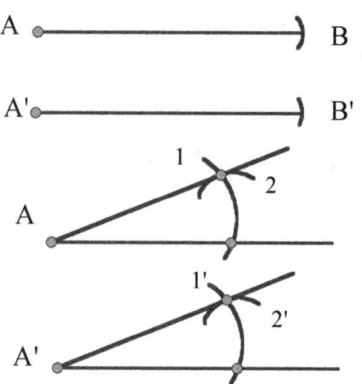

(2) Bisect a line segment or an angle.

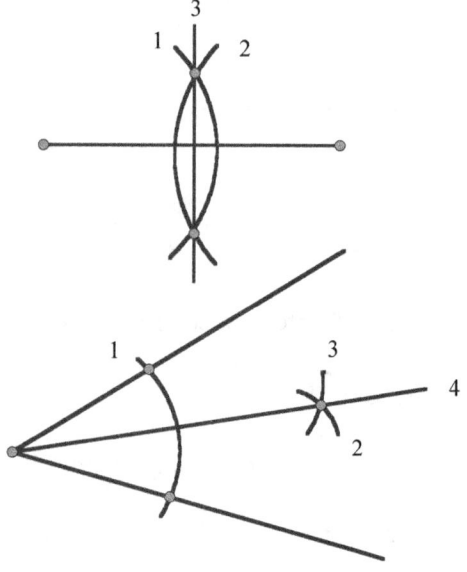

(3) Through a point draw a line ⊥ or ∥ to a given line.

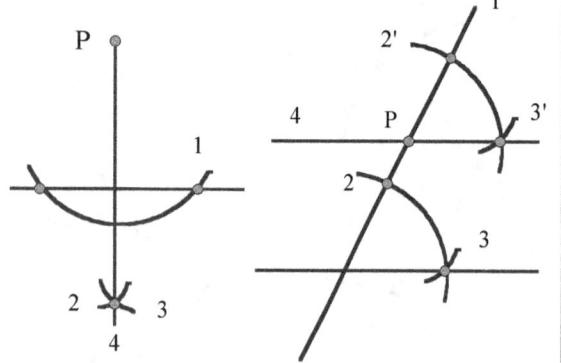

2. Five Fundamental Loci

(1) The locus of points equidistant from a given point.
(2) The locus of points equidistant from two given points.
(3) The locus of points equidistant from two sides of a given angle.
(4) The locus of points equidistant from a given line.
(5) The locus of points equidistant from two given parallel lines.

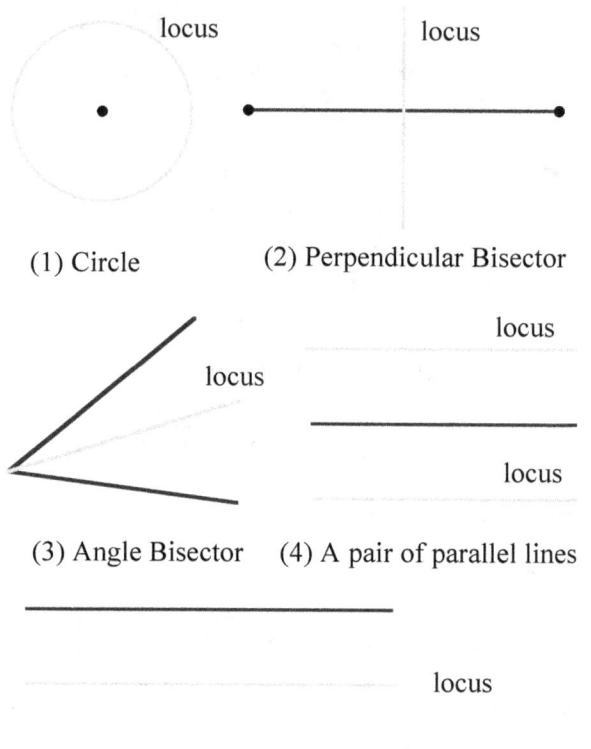

(1) Circle   (2) Perpendicular Bisector

(3) Angle Bisector   (4) A pair of parallel lines

(5) A parallel line midway between the given lines

3. Compound Loci

Find the points of intersection of different loci.
e.g. How many points are 2 units from a given line and 3 units from a given point on the given line ?

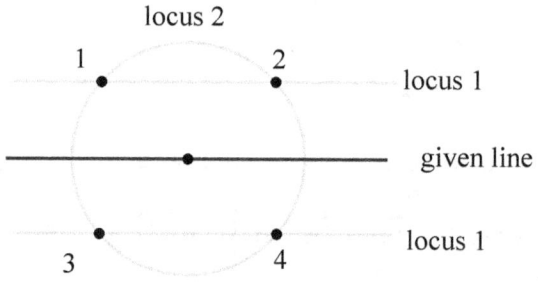

There are four points of intersection.

4. Equations of Loci

(1) The center-radius equation of a circle with radius r and center (h, k)

$$(x - h)^2 + (y - k)^2 = r^2$$

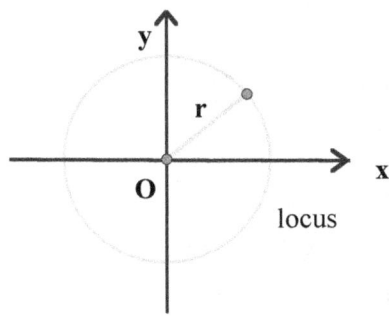

locus

$$x^2 + y^2 = r^2$$

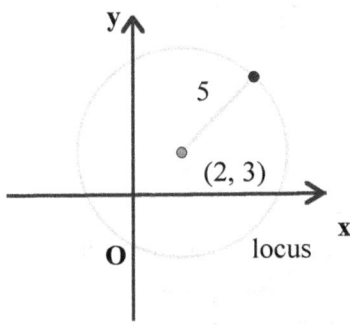

locus

$$(x - 2)^2 + (y - 3)^2 = 5^2$$

(2) The equation of the perpendicular bisector

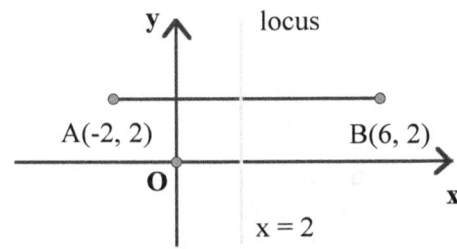

locus

$x = 2$ is the equation of the perpendicular bisector of $\overline{AB}$.

Find the equation of the locus of points equidistant from points A(-2, 2) and B(4, -2).

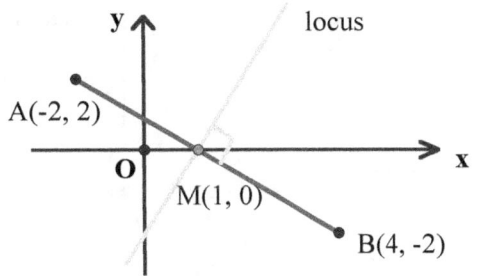

locus

Find the midpoint of $\overline{AB}$:
$$M(\bar{x}, \bar{y}) = M(\frac{-2 + 4}{2}, \frac{2 + (-2)}{2})$$
$$= M(1, 0)$$

Find the slope of $\overline{AB}$:
$$m_1 = \frac{-2 - 2}{4 - (-2)} = \frac{-4}{6} = -\frac{2}{3}$$

the slope of the perpendicular line:
$$m_2 = -\frac{1}{m_1} = \frac{3}{2}$$

the equation of the perpendicular bisector:

the slope is $\frac{3}{2}$ and passing through midpoint (1, 0)

the point-slope form: $y - y_1 = m(x - x_1)$
$$y - 0 = \frac{3}{2}(x - 1), \quad \text{or}$$

slope-intercept form: $y = \frac{3}{2}x - \frac{3}{2}$

(3) The equation of the angle bisector.

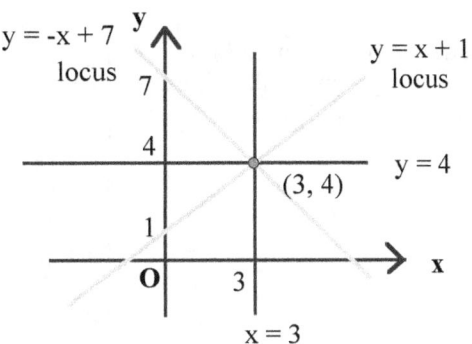

locus

$y = x + 1$ and $y = -x + 7$ are the loci of the points equidistant from lines $x = 3$ and $y = 4$.

Hint: $m = \pm 1$ and passing through point (3, 4)

(4) The equations of the locus of points equidistant from a given line.

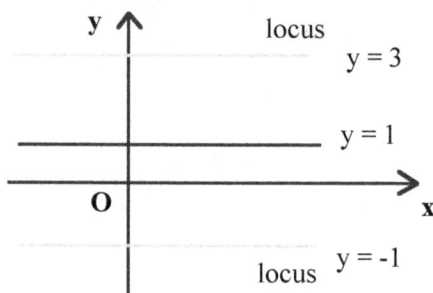

$y = -1$ and $y = 3$ are the equations of the parallel lines 2 units from the given line $y = 1$

(5) The equation of the locus of points equidistant from two given parallel lines.

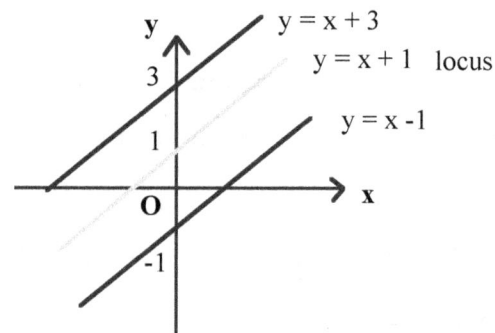

The y-intercept of the locus is equal to

$$\frac{-1 + 3}{2} = 1$$

$y = x + 1$ is the equation of the parallel line midway between the two given parallel lines. They have the same slope 1 in this example.

(6) The points of intersection of compound loci:

Find the solution to the quadratic-linear system of equations.

e.g. Solve the quadratic-linear system of equations
$$y = x^2 + 1$$
$$y = x + 3$$

Method 1: Solve graphically

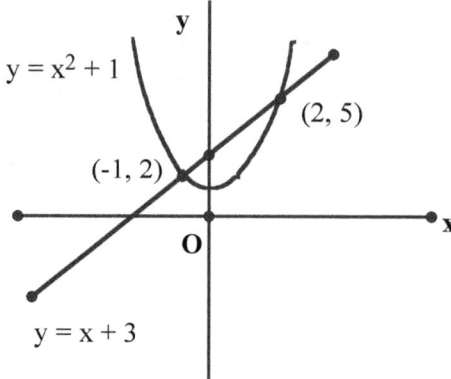

To graph the parabola $y = x^2 + 1$, properly choose 7 points around the turning point:

x	-3	-2	-1	0	1	2	3
y	10	5	2	1	2	5	10

The points of intersection $(-1, 2)$ and $(2, 5)$ are the solution to the quadratic-linear system of equations.

Method 2: Solve algebraically
$$y = x^2 + 1$$
$$y = x + 3$$
$x^2 + 1 = x + 3$ substitution
$x^2 - x - 2 = 0$ make the right side 0
$(x - 2)(x + 1) = 0$ factor the left side
$x - 2 = 0$ or $x + 1 = 0$
$x = 2$ or $x = -1$
$y = x + 3 = 5$ $y = x + 3 = 2$
Solution: $\{(2, 5), (-1, 2)\}$

X. TRANSFORMATION

1. Transformation Rules

(1). Line Reflection:

$P(x, y)$ _____rx-axis_____ $P'(x, -y)$
$P(x, y)$ _____ry-axis_____ $P'(-x, y)$
$P(x, y)$ _____ry = x_____ $P'(y, x)$
$P(x, y)$ _____ry = - x_____ $P'(- y, -x)$

(2). Point Reflection:

$P(x, y)$ _____ro_____ $P'(-x, -y)$

(3). Translation:

$P(x, y)$ _____$T_{a, b}$_____ $P'(x + a, y + b)$

(4). Rotation:

$P(x, y)$ _____$R\,90°$_____ $P'(-y, x)$
$P(x, y)$ _____$R\,180°$_____ $P'(-x, -y)$
$P(x, y)$ _____$R\,-90°$_____ $P'(y, -x)$

(5). Dilation:

$P(x, y)$ _____$D\,k$_____ $P'(kx, ky)$

Only dilation enlarges or reduces the size of the image, which is similar to the original.
The image of other transformations is congruent to the original.

It is not practical to memorize these rules. One should be able to derive these rules through the drawings.

e.g. Find the rules for ry-axis and R -90°.

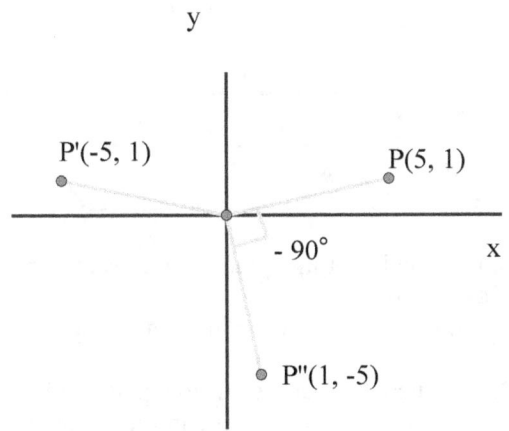

$P(5, 1)$	_____ry-axis_____	$P'(-5, 1)$
$P(x, y)$	_____ry-axis_____	$P'(-x, y)$
$P(5, 1)$	_____$R\,-90°$_____	$P''(1, -5)$
$P(x, y)$	_____$R\,-90°$_____	$P''(y, -x)$

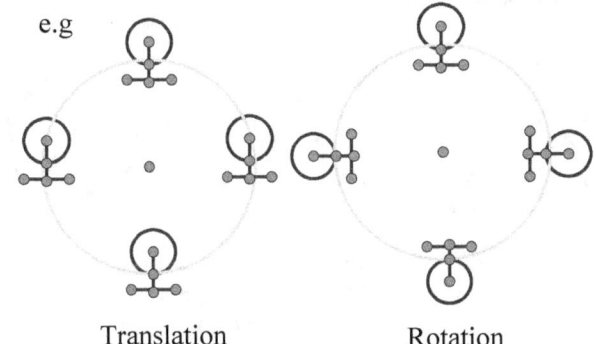

Translation   Rotation

2. Composition of Transformations

e.g. rx-axis ∘ ry-axis (x, y) = rx-axis $(-x, y)$ = $(-x, -y)$
    We can see   rx-axis ∘ ry-axis = ro

e.g. $D_4 ∘ T_{3, 0}(x, y) = D_4(x + 3, y) = (4x + 12, 4y)$

or $(x, y) \xrightarrow{T_{3, 0}} (x + 3, y) \xrightarrow{D_4} (4x + 12, 4y)$

Glide Reflection:

Glide Reflection is a special composition of reflections and translations: $T_{a, 0} ∘ r_{x\text{-axis}}$

e.g. Find the image of $P(1, 2)$ under the glide reflection of $T_{2,0} ∘ r_{x\text{-axis}}$

$P(1, 2)$ _____rx-axis_____ $P'(1, -2)$ _____$T_{2,0}$_____ $P''(3, -2)$

3. Functions under a Transformation

$y = f(x)$ _____$T_{a,b}$_____ $y = f(x - a) + b$

$y = f(x)$ _____rx-axis_____ $y = - f(x)$

The transformation rules for functions are different from the transformation rules for images

e.g. $y = x^2$ _____$T_{5,2}$_____ $y = (x - 5)^2 + 2$
    $y = x^2$ _____rx-axis_____ $y = - x^2$

4. Symmetry

(1). Line Symmetry

e.g.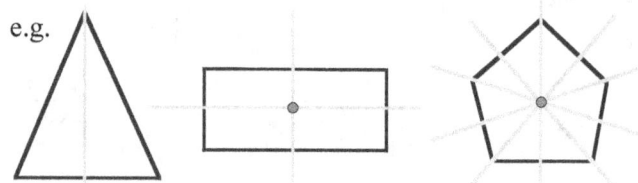

Isosceles triangle has 1 axis of symmetry;
Rectangle has 2 axes of symmetry;
Regular pentagon has 5 axes of symmetry.

(2). Point Symmetry

e.g.

(3). Rotational Symmetry

e.g.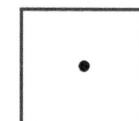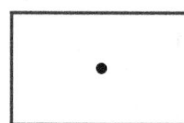

Equilateral triangle has 120° rotational symmetry;
Square has 90° rotational symmetry;
Rectangle has 180° rotational symmetry;
Regular pentagon has 72° rotational symmetry;
Regular hexagon has 60° rotational symmetry.

5. Isometry and Orientation

Isometry: A transformation that preserves the distance.
Only dilation is not isometry.

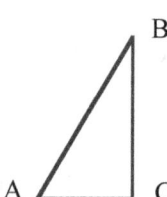 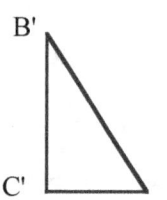

From A ---> B ---> C is **clockwise orientation**.
From A' ---> B' ---> C' is **counterclockwise orientation**.

Direct Isometry: An isometry preserves orientation;
Opposite Isometry: An isometry changes orientation.

e.g. Line Reflection: Opposite Isometry
Point Reflection: Direct Isometry
Rotation: Direct Isometry
Translation: Direct Isometry
Dilation: Changes size --- Not an isometry,
but preserves orientation

The composition of a direct isometry and an opposite isometry is an opposite isometry.
The composition of two opposite isometries is a direct isometry.

XI. SOLID GEOMETRY

1. To determine a plane

(1). Three noncollinear points determine a plane.

(2). Two intersecting lines determine a plane.

(3). Two parallel lines determine a plane.

(4). Skew lines are neither parallel nor intersecting, they are not in a same plane --- not coplanar.

e.g. In a triangular right prism

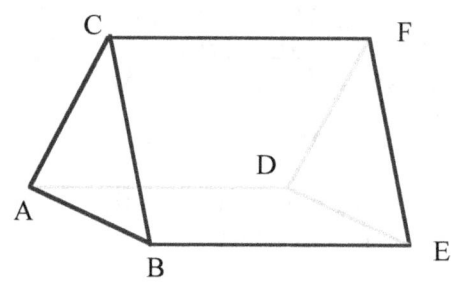

$\overline{AB}$ ∥ $\overline{DE}$, $\overline{BC}$ ∥ $\overline{EF}$, $\overline{CA}$ ∥ $\overline{FD}$;
$\overline{AD}$ ∥ $\overline{BE}$ ∥ $\overline{CF}$;
$\overline{AB}$ and $\overline{CF}$, $\overline{BC}$ and $\overline{AD}$, $\overline{CA}$ and $\overline{BE}$,
$\overline{AB}$ and $\overline{FD}$, $\overline{AB}$ and $\overline{FE}$ etc. are pairs of skew lines.

2. A line perpendicular to a plane

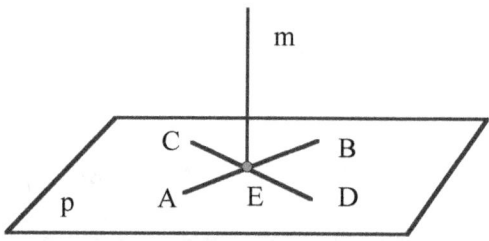

(1). If a line is not in a plane, it intersects a plane in exactly one point.
e.g. line m intersects plane p at point E

(2). If a line is perpendicular to a plane, it is perpendicular to each line in the plane through the point of intersection.
e.g. $m \perp \overleftrightarrow{AB}$, $m \perp \overleftrightarrow{CD}$

(3). Through a given point (on the plane or not on the plane), there is one and only one line perpendicular to the given plane.

(4). Two lines perpendicular to a same plane are parallel and coplanar.

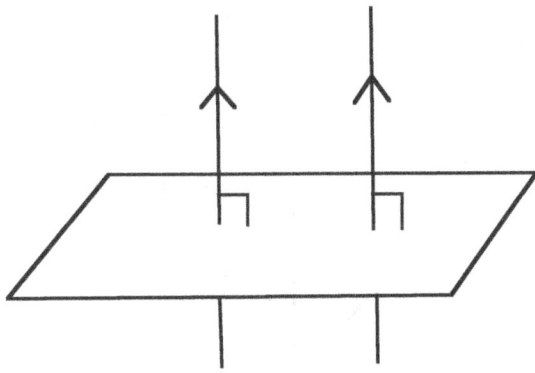

3. Perpendicular Planes

(1). Two perpendicular planes intersect to form a right dihedral angle.

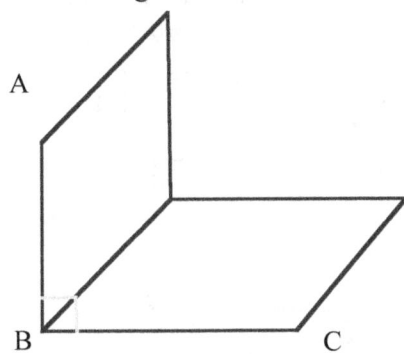

(2). If plane a and plane b are both perpendicular to plane c , then their line of intersection m is perpendicular to plane c.

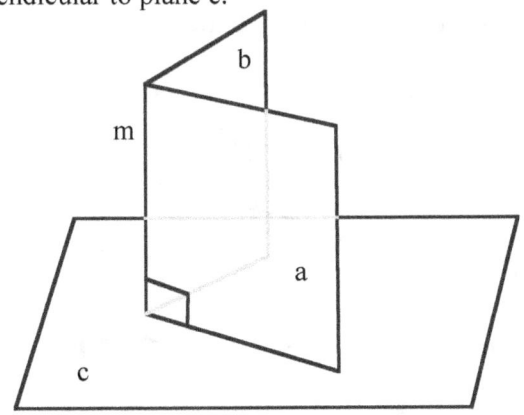

(3). If a plane contains a line perpendicular to another plane, then these two planes are perpendicular.

4. Parallel Planes

(1). Parallel planes are everywhere equidistant.

(2). If two planes are perpendicular to a same line, then they are parallel.

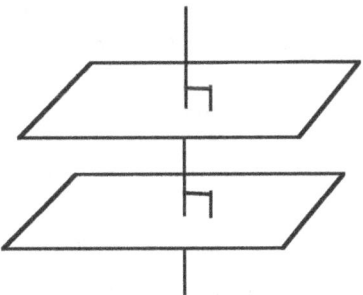

(3). If a plane intersects two parallel planes, then the intersection is two parallel lines.

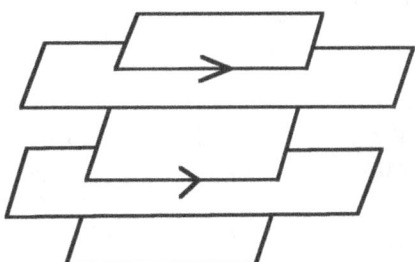

XII. INDIRECT PROOF

(1) Assume that the opposite of the conclusion is true.
(2) Show that the assumption contradicts a known fact.
(3) Since the assumption is false, the conclusion is true.

e.g.

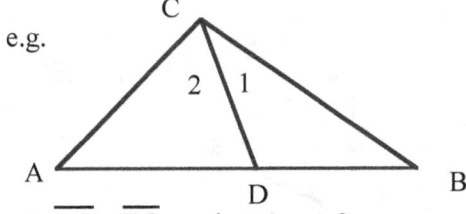

Given: $\overline{AC} \cong \overline{BC}$ and $\angle 1 \neq \angle 2$
Prove: $\overline{CD}$ is not a median

(1) $\overline{AC} \cong \overline{BC}$, $\angle 1 \neq \angle 2$ Given
(2) $\overline{CD}$ is a median   Assumed
(3) D is the midpoint   Def. of median
(4) $\overline{AD} \cong \overline{BD}$ Def. of midpoint
(5) $\overline{CD} \cong \overline{CD}$ Reflexive Property
(6) $\triangle ACD \cong \triangle BCD$ S.S.S $\cong$
(7) $\angle 1 \cong \angle 2$ CPCTC
(8) $\overline{CD}$ is not a median   Contradiction in (7) and (1)

XIII. GEOMETRIC MEASUREMENTS

1 yd = 3 ft , 1 ft = 12 in , 1 mile = 5280 ft
1 m = 100 cm , 1 m = 1000 mm

1. Circle

Circumference $C = 2\pi r = \pi d$ r: radius   d: diameter
Area $A = \pi r^2$
e.g. When r is doubled, C is doubled and A increases 4 times.

2. Square

Perimeter $P = 4s$ s: length of the side
Area $A = s^2$

3. Rectangle

Perimeter $P = 2l + 2w$ l: length   w: width
Area $A = l \cdot w$

4. Parallelogram

Perimeter P = sum of the 4 sides   b: base   h: height
Area $A = b \cdot h$

5. Trapezoid

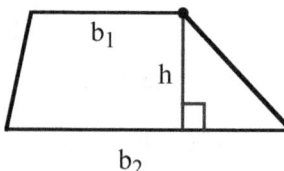

P = sum of 4 sides

$$A = \frac{b_1 + b_2}{2} \cdot h$$

6. Rhombus

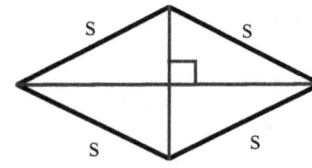

$P = 4s$

$$A = \frac{1}{2} \cdot d_1 \cdot d_2$$

d_1 and d_2 are diagonals

7. Triangle

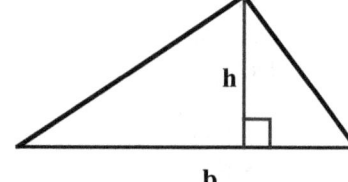

P = sum of 3 sides

$$A = \frac{1}{2} \cdot b \cdot h$$

8. Right Prism

Volume $V = Bh$ B: Base Area, h: height
Surface Area = 2 Base Areas + Lateral Areas

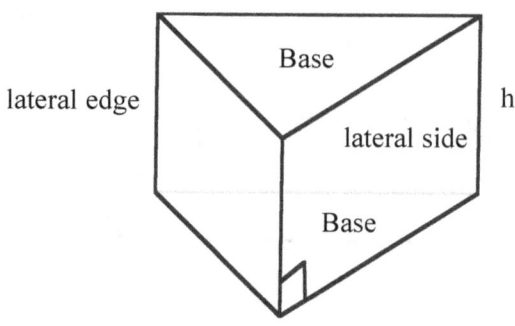

All of the lateral edges are congruent and parallel.

Special cases:

Rectangular Prism
Volume $V = lwh$ l: length, w: width, h: height
Surface Area $SA = 2wl + 2wh + 2hl$

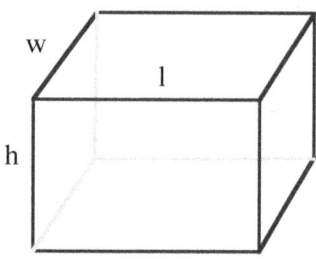

Cube
Volume $V = s^3$ s : length of the side
Surface Area $SA = 6s^2$

9. Right Circular Cylinder

Volume $V = Bh$ B: area of the circular base πr^2
               h: height
Lateral Area $L = 2\pi rh$

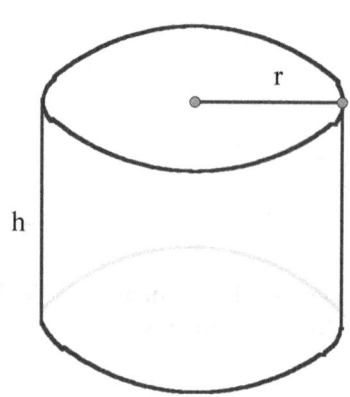

10. Pyramid

Volume $V = \dfrac{1}{3}Bh$ B: Base Area, h: height

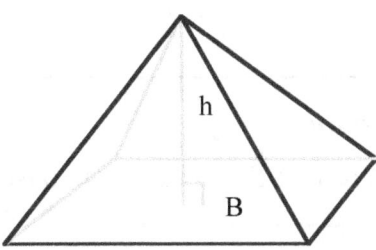

The base of a regular pyramid is a regular polygon. The lateral sides of a regular pyramid are congruent isosceles triangles.

11. Right Circular Cone

Volume $V = \dfrac{1}{3}Bh$ B: Base Area $= \pi r^2$, h: height

Lateral Area $L = \pi rl$ l: slant height

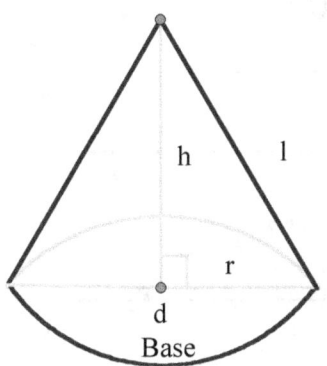

e.g.
A right circular cone has a diameter of 16 and a height of 18. Find the volume.
(a). express the answer in terms of π .
(b). express the answer to the nearest tenth

$$r = \frac{d}{2} = \frac{16}{2} = 8$$

$$B = \pi r^2 = \pi \cdot 8^2 = 64\pi$$

$$V = \frac{1}{3}Bh = \frac{1}{3} \cdot 64\pi \cdot 18$$

(a). $V = 384\pi$

(b). $V = 1206.4$

12. Sphere

Volume $V = \dfrac{4}{3}\pi r^3$

Surface Area $SA = 4\pi r^2$

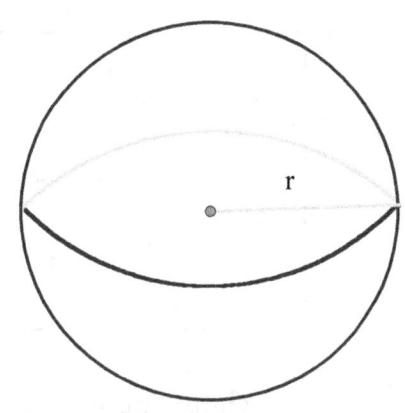

Geometry Reference Sheet

Volume	Cylinder	$V = Bh$ where B is the area of the base
	Pyramid	$V = \frac{1}{3}Bh$ where B is the area of the base
	Right Circular Cone	$V = \frac{1}{3}Bh$ where B is the area of the base
	Sphere	$V = \frac{4}{3}\pi r^3$

Lateral Area (L)	Right Circular Cylinder	$L = 2\pi rh$
	Right Circular Cone	$L = \pi rl$ where l is the slant height

Surface Area	Sphere	$SA = 4\pi r^2$

PART 1. Algebra

I. ALGEBRAIC EXPRESSIONS, EQUATIONS, AND INEQUALITIES

1. Factoring Polynomials

Find Common Factors:
$$3x^2 + 6x = 3x(x + 2)$$
$$2y^3 - 4y^2 + 2y = 2y(y^2 - 2y + 1)$$
The Difference of Two Squares:
$$a^2 - b^2 = (a + b)(a - b)$$
$$4y^2 - 25 = (2y + 5)(2y - 5)$$
Trinomial:
$$x^2 + 2x - 15 = (x + 5)(x - 3)$$
Here $5 \cdot (-3) = -15$ and $5 + (-3) = 2$

Four-term Polynomial:
If the product of the first and last terms is equal to the product of the two middle terms, the polynomial can be grouped.
e.g.
$$3x^3 - 6x^2 + 2x - 4$$
$$= 3x^2(x - 2) + 2(x - 2)$$
$$= (3x^2 + 2)(x - 2)$$
Factor Completely:
$$2x^3 - 14x^2 + 20x$$
$$= 2x(x^2 - 7x + 10) = 2x(x - 2)(x - 5)$$

2. Quadratic Equations

(1). Use Factoring
e.g. $x^2 - 10 = 3x$
$x^2 - 3x - 10 = 0$ set one side equal to zero
$(x + 2)(x - 5) = 0$ factor the trinomial
$x + 2 = 0$ or $x - 5 = 0$
$x = -2$ or $x = 5$
solution set $\{-2, 5\}$

(2). Complete the Square
$$x^2 + bx + c = 0$$
$$x^2 + bx = -c$$
$$x^2 + bx + \left(\frac{b}{2}\right)^2 = -c + \left(\frac{b}{2}\right)^2$$
$$\left(x + \frac{b}{2}\right)^2 = -c + \left(\frac{b}{2}\right)^2$$
e.g. $x^2 - 8x + 3 = 0$
$$x^2 - 8x + (-4)^2 = -3 + (-4)^2$$
$$(x - 4)^2 = 13$$
$$x - 4 = \pm\sqrt{13}$$
$$x = 4 \pm \sqrt{13}$$

(3). Quadratic Formula

$$ax^2 + bx + c = 0 \quad \text{where } a \neq 0$$
$$x = \frac{-b \pm \sqrt{b^2 - 4ac}}{2a}$$

e.g. $2x^2 - 4x + 1 = 0$
$a = 2$, $b = -4$, $c = 1$
$$x = \frac{4 \pm \sqrt{(-4)^2 - 4(2)(1)}}{2(2)}$$
$$= \frac{4 \pm \sqrt{8}}{4} = \frac{4 \pm 2\sqrt{2}}{4}$$
$$= \frac{2 \pm \sqrt{2}}{2}$$

Quadratic Inequalities

The solution set of a quadratic inequality is in the form of

(1) $x_1 < x < x_2$ where x_1 and x_2 are the roots
or (2) $x < x_1$ or $x > x_2$ $x_1 < x_2$

Use the quadratic inequality to test a value from each of the three regions.

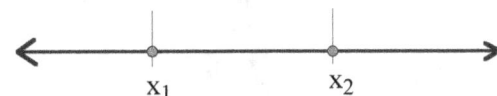

e.g. $x^2 - 7x + 10 < 0$
Solve the corresponding quadratic equation
$$x^2 - 7x + 10 = 0$$
$$x_1 = 2, \quad x_2 = 5$$

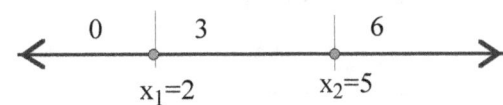

test $x = 0$ (false), $x = 3$ (true), $x = 6$ (false);
therefore the solution set is $2 < x < 5$

e.g. $x^2 - 7x + 10 > 0$
$$x_1 = 2, \quad x_2 = 5$$

test $x = 0$ (true), $x = 3$ (false), $x = 6$ (true);
therefore the solution set is $x < 2$ or $x > 5$

Quadratic-Linear System

(1). Algebraic Solution:

e.g.
$$y = x^2 - 8 \quad\quad (1)$$
$$y + 5 = 2x \quad\quad (2)$$

From Eq. (2) $y = 2x - 5$ (3)

Substitute y by $2x - 5$ in Eq.(1):
$$2x - 5 = x^2 - 8$$
$$x^2 - 2x - 3 = 0$$
$$(x - 3)(x + 1) = 0$$

$x - 3 = 0$	$x + 1 = 0$
$x = 3$	$x = -1$
$y = 2(3) - 5 = 1$	$y = 2(-1) - 5 = -7$

Solution: $\{(3, 1), (-1, -7)\}$

(2). Graphic Solution:

e.g.
$$y = x^2 - 8$$
$$y + 5 = 2x$$

rewrite in the slope-intercept form
$$y = 2x - 5$$

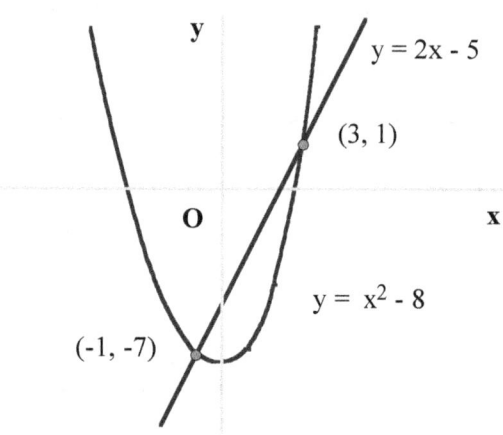

The solution of the system of equations are the intersection points (3, 1) and (-1, -7)

*Refer to **PART 4. Graphing Calculator**
Use the graphing calculator to list the table of the points and show the graphs.

3. Rational Expressions and Operations

Denominator cannot be zero.

e.g. $\dfrac{x}{x - 2}$ when $x = 2$, $x - 2 = 0$, undefined

(1). Simplify: $\dfrac{2x^3}{x^2 - x - 12} \cdot \dfrac{x^2 - 16}{6x}$

$= \dfrac{2x^3(x + 4)(x - 4)}{(x + 3)(x - 4)\bullet 6x}$ factor the numerator

and the denominator first;

$= \dfrac{x^2(x + 4)}{3(x + 3)}$ cancel out common factors in

the numerator and the denominator

(2). Divide: $\dfrac{\dfrac{2x}{x + 1}}{1 - \dfrac{2x}{x + 1}} = \dfrac{\dfrac{2x}{x + 1}}{\dfrac{x + 1 - 2x}{x + 1}} = \dfrac{\dfrac{2x}{x + 1}}{\dfrac{1 - x}{x + 1}}$

$= \dfrac{2x}{x + 1} \bullet \dfrac{x + 1}{1 - x} = \dfrac{2x}{1 - x}$

(3). Combine: $\dfrac{1}{x + 1} + \dfrac{x - 1}{2}$ LCD $= 2(x + 1)$

$= \dfrac{2\bullet 1}{2(x + 1)} + \dfrac{(x - 1)\bullet(x + 1)}{2(x + 1)}$

$= \dfrac{2 + x^2 - 1}{2(x + 1)} = \dfrac{x^2 + 1}{2(x + 1)}$

Rational Equations

(1). Use cross-multiplication for proportions:

e.g. $\dfrac{x + 2}{x - 3} = \dfrac{x}{4}$

$4(x + 2) = x(x - 3)$ cross-multiply
$4x + 8 = x^2 - 3x$
$x^2 - 7x - 8 = 0$
$(x - 8)(x + 1) = 0$
$x = 8$ or $x = -1$ must check: Both are solutions.

(2). Use LCD:

e.g. $\dfrac{2}{x^2 - x} = \dfrac{2}{x - 1} + 1$ LCD $= x(x - 1)$

$2 = 2x + x(x - 1)$ multiply LCD on both sides
$2 = 2x + x^2 - x$
$x^2 + x - 2 = 0$
$(x + 2)(x - 1) = 0$
$x = -2$ or $x = 1$
check: $x = -2$ ($x = 1$ undefined)

Algebra 2 and Trigonometry Review

Rational Inequalities

e.g. $\dfrac{x+2}{x-3} > \dfrac{x}{4}$

Step 1. Solve the corresponding equation

$$\dfrac{x+2}{x-3} = \dfrac{x}{4}$$

$$x = 8 \ \text{ or } \ x = -1$$

Step 2. Find the undefined values of x

$$x - 3 = 0$$
$$x = 3$$

Step 3. Divide the number line by these values

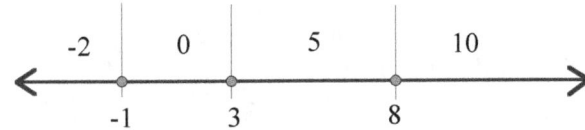

test x = -2 (true), x = 0 (false), x = 5 (true)
      x = 10 (false)
thus the solution set is $x < -1$ or $3 < x < 8$
   or $(-\infty, -1) \cup (3, 8)$

Note: The true and false intervals always alternate.

4. Radicals

$$\sqrt{a \cdot b} = \sqrt{a} \cdot \sqrt{b} \qquad a \geq 0, b \geq 0$$

$$\sqrt{\dfrac{a}{b}} = \dfrac{\sqrt{a}}{\sqrt{b}} \qquad a \geq 0, b > 0$$

Simplify: $\sqrt{75} = \sqrt{25 \cdot 3} = 5\sqrt{3}$
$ \sqrt{300} = \sqrt{100 \cdot 3} = 10\sqrt{3}$

Combine: $5\sqrt{x} + 3\sqrt{x} = 8\sqrt{x}$
$ \sqrt{18} - 4\sqrt{2} = \sqrt{9 \cdot 2} - 4\sqrt{2} = 3\sqrt{2} - 4\sqrt{2} = -\sqrt{2}$

Multiply: $2\sqrt{3} \cdot 4\sqrt{5} = 2 \cdot 4\sqrt{3 \cdot 5} = 8\sqrt{15}$
$ 3\sqrt{2} \cdot 7\sqrt{2} = 3 \cdot 7\sqrt{2 \cdot 2} = 21 \cdot 2 = 42$

Divide: $\dfrac{4\sqrt{6}}{2\sqrt{3}} = \dfrac{4}{2}\sqrt{\dfrac{6}{3}} = 2\sqrt{2}$

Rationalize: $\dfrac{1}{\sqrt{3}} = \dfrac{1}{\sqrt{3}} \cdot \dfrac{\sqrt{3}}{\sqrt{3}} = \dfrac{\sqrt{3}}{3}$

Exponents

$a^0 = 1 \ \ (a \neq 0) \qquad 5^0 = 1, (-5)^0 = 1, 1.2^0 = 1$

$x^{-n} = \dfrac{1}{x^n} \ \ (x \neq 0) \qquad 5^{-2} = \dfrac{1}{5^2} = \dfrac{1}{25}$

$x^a \cdot x^b = x^{a+b} \qquad\qquad 5^2 \cdot 5^3 = 5^{2+3} = 5^5$

$\dfrac{x^a}{x^b} = x^{a-b} \qquad \dfrac{8xy^3}{2xy} = \dfrac{8}{2} \cdot \dfrac{x}{x} \cdot \dfrac{y^3}{y} = 4y^2$

$(x^a)^b = x^{a \cdot b} \qquad\qquad (5^2)^3 = 5^{2 \cdot 3} = 5^6$

Rational Exponents

$$x^{\frac{a}{b}} = \sqrt[b]{x^a} \qquad x^{\frac{a}{b}} = \left(\sqrt[b]{x}\right)^a \qquad x \geq 0$$

e.g. $x^{\frac{1}{2}} = \sqrt{x} \ , \quad x^{\frac{2}{3}} = \sqrt[3]{x^2}$

e.g. $125^{\frac{2}{3}} = \left(\sqrt[3]{125}\right)^2 = 5^2 = 25$

e.g. $\dfrac{3^{\frac{1}{3}}}{3^{-\frac{2}{3}}} = 3^{\frac{1}{3} - \left(-\frac{2}{3}\right)} = 3^1 = 3$

Rationalize the Denominator

e.g. $\dfrac{2}{4 + \sqrt{11}}$

$= \dfrac{2}{4 + \sqrt{11}} \cdot \dfrac{4 - \sqrt{11}}{4 - \sqrt{11}}$ multiply its conjugate

$= \dfrac{2(4 - \sqrt{11})}{16 - 11} = \dfrac{8 - 2\sqrt{11}}{5}$

e.g. $\dfrac{1}{\sqrt{3}} + \dfrac{1}{\sqrt{2}}$

$= \dfrac{1}{\sqrt{3}} \cdot \dfrac{\sqrt{3}}{\sqrt{3}} + \dfrac{1}{\sqrt{2}} \cdot \dfrac{\sqrt{2}}{\sqrt{2}}$ rationalize first

$= \dfrac{\sqrt{3}}{3} + \dfrac{\sqrt{2}}{2}$ $\text{LCD} = 6$

$= \dfrac{2\sqrt{3} + 3\sqrt{2}}{6}$

Algebra 2 and Trigonometry Review

Radical Equations

e.g. $x = 1 + \sqrt{x+5}$

$x - 1 = \sqrt{x+5}$ isolate the radical

$(x-1)^2 = x + 5$ square both sides

$x^2 - 2x + 1 = x + 5$

$x^2 - 3x - 4 = 0$

$x = 4$ or $x = -1$

must check: $x = 4$ ($x = -1$ rejected)

e.g. $x^{\frac{3}{2}} + 1 = 9$

$x^{\frac{3}{2}} = 8$ isolate the radical

$(x^{\frac{3}{2}})^{\frac{2}{3}} = 8^{\frac{2}{3}}$

$x = 4$

must check: $x = 4$

5. Absolute Value Equations

e.g. $|4 - x| = 3x$

remove the absolute value symbol

$4 - x = 3x$	$-(4-x) = 3x$
$x = 1$	$x = -2$
check: $\lvert 4-1\rvert = 3 \cdot 1$	check: $\lvert 4 - (-2)\rvert \neq 3 \cdot (-2)$
(true)	(false)

e.g. $|2x + 5| - 4 = x$

$(2x+5) - 4 = x$	$-(2x+5) - 4 = x$
$x = -1$	$x = -3$
(true)	(true)

Absolute Value Inequality

(1) If $|x| < k$ where $k > 0$

then $-k < x < k$

e.g. $|2x + 3| < 7$

$-7 < 2x + 3 < 7$

$-7 < 2x + 3$	$2x + 3 < 7$
$-10 < 2x$	$2x < 4$
$-5 < x$	$x < 2$

solution: $-5 < x < 2$

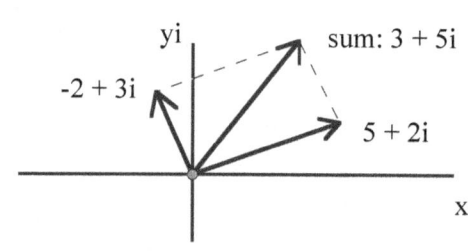

-5 2

(2) If $|x| > k$ where $k > 0$

then $x < -k$ or $x > k$

e.g. $|10 - 2x| - 2 \geq 0$

           $|10 - 2x| \geq 2$

$10 - 2x \leq -2$ or $10 - 2x \geq 2$

$-2x \leq -12$ or $-2x \geq -8$ divided by a negative number

$x \geq 6$ or $x \leq 4$ inequality sign reversed

4   6

6. Complex Numbers

$a + bi$ where a, b real numbers, i imaginary unit.

$a + bi = c + di$ if and only if $a = c$ and $b = d$.

e.g.

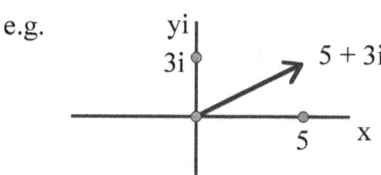

The magnitude (length) of the vector of a complex number:

e.g $|5 + 3i| = \sqrt{5^2 + 3^2} = \sqrt{34}$

Def. of the Imaginary Number:

$i = \sqrt{-1}$, $i^2 = -1$

$i^0 = 1$, $i^1 = i$, $i^2 = -1$, $i^3 = -i$

$i^{4n} = 1$, $i^{4n+1} = i$, $i^{4n+2} = -1$, $i^{4n+3} = -i$

e.g. $i^{82} = i^{4 \cdot 20 + 2} = i^2 = -1$

e.g. $\sqrt{-16} + \sqrt{-9} = 4i + 3i = 7i$

e.g. $\sqrt{-16} \cdot \sqrt{-9} = 4i \cdot 3i = 12\, i^2 = -12$

(Note: $\sqrt{-16} \cdot \sqrt{-9} \neq \sqrt{(-16)(-9)} = \sqrt{144} = 12$)

Addition: $(5 + 2i) + (-2 + 3i)$

represented algebraically:

$= (5 - 2) + (2i + 3i) = 3 + 5i$

represented graphically:

Multiplication: $(3 + i)(2 - 2i)$
$$= 6 - 6i + 2i - 2i^2 = 6 - 4i + 2 = 8 - 4i$$

Conjugates:
$a + bi$ and $a - bi$ are conjugates.
The sum or the product of two conjugates is a real number.
$(a + bi) + (a - bi) = 2a$
$(a + bi)(a - bi) = a^2 - (bi)^2 = a^2 - b^2 i^2 = a^2 + b^2$

Division: $\dfrac{8 + i}{2 - i}$

$= \dfrac{(8 + i)(2 + i)}{(2 - i)(2 + i)}$ $2 + i$ is the conjugate of $2 - i$

$= \dfrac{16 + 8i + 2i + i^2}{4 - i^2} = \dfrac{15 + 10i}{5} = 3 + 2i$

Multiplicative Inverse:
e.g. Write the multiplicative inverse of $2 + 4i$ in $a + bi$ form.
$$\frac{1}{2 + 4i} = \frac{1}{2 + 4i} \cdot \frac{2 - 4i}{2 - 4i} = \frac{2 - 4i}{4 + 16} = \frac{2 - 4i}{20} = \frac{1}{10} - \frac{1}{5}i$$

e.g. The multiplicative inverse of $5i$ is $\dfrac{1}{5i} = \dfrac{1}{5i} \cdot \dfrac{i}{i} = -\dfrac{i}{5}$

7. Roots of the Quadratic Equations

$ax^2 + bx + c = 0$ where $a \neq 0$

The sum of the roots $x_1 + x_2 = -\dfrac{b}{a}$

The product of the roots $x_1 \cdot x_2 = \dfrac{c}{a}$

e.g. $x_1 + x_2 = 5$ and $x_1 \cdot x_2 = 6$
    Write the quadratic equation.
$$5 = -\frac{b}{a} , \quad 6 = \frac{c}{a}$$
$$a = 1 , b = -5 , c = 6$$
$$x^2 - 5x + 6 = 0$$

$b^2 - 4ac$ is called discriminant Δ
$\Delta > 0$ two unequal real roots (two x intercepts)
$\Delta = 0$ two equal real roots (one x intercept)
$\Delta < 0$ no real roots (no x intercept)
when $\Delta < 0$, it has two imaginary roots in the form of
conjugates: $x_1 = a + bi$ and $x_2 = a - bi$
e.g. $x^2 + 2x + 2 = 0$ $x_1 = -1 + i$, $x_2 = -1 - i$

8. Polynomial Equations of Higher Degrees

A polynomial equation of degree n has n roots which are real, imaginary, or both.
The real roots are the x-intercepts and the imaginary roots are paired complex conjugates.

e.g $x^3 - 3x^2 + 4x - 12 = 0$
      $x^2(x - 3) + 4(x - 3) = 0$
      $(x - 3)(x^2 + 4) = 0$

$x - 3 = 0$	$x^2 + 4 = 0$
$x = 3$	$x^2 = -4$
	$x = \pm\sqrt{-4}$
	$x = \pm 2i$

Solution: $\{ 3, -2i, 2i \}$

e.g. $x^4 + 3x^2 = 4$
      $x^4 + 3x^2 - 4 = 0$
      $(x^2 + 4)(x^2 - 1) = 0$

$x^2 + 4 = 0$	$x^2 - 1 = 0$
$x = \pm 2i$	$(x + 1)(x - 1) = 0$
	$x = -1 , x = 1$

Solution set: $\{ -1, 1, -2i, 2i \}$

9. Sequence and Series

(1). A sequence is a special type of function:
    $a_n = f(n)$ where $\{ n: 1, 2, 3, 4, 5, \bullet \bullet \bullet \}$

Recursive Definition:
    In some sequences, except the first one, any term can be defined by its previous terms.

e.g.   sequence $\{2, 4, 8, 16, \bullet \bullet \bullet \}$
          $a_n = 2^n$
    use recursive definition:
       $a_n = 2 a_{n-1}$ for $n > 1$

A **series** is the sum of the sequence.

e.g. $S_5 = 2 + 4 + 8 + 16 + 32$

Algebra 2 and Trigonometry Review

(2). Sigma Notation $\sum$

e.g.
$$\sum_{i=1}^{5} i = 1 + 2 + 3 + 4 + 5$$

here i is the index, lower limit = 1, upper limit = 5

e.g.
$$\sum_{n=1}^{5} (2n - 1) = 1 + 3 + 5 + 7 + 9$$

e.g.
$$\sum_{k=3}^{7} k^2 = 3^2 + 4^2 + 5^2 + 6^2 + 7^2$$

(3). Arithmetic Sequences

$a_n = a_1 + (n - 1)\,d$ d is the common difference

or $a_n = a_{n-1} + d$ recursive definition

e.g. { 5, 10, 15, 20, 25 }

e.g. Find the missing terms between $a_1 = 5$ and $a_6 = 15$ in an arithmetic sequence.

$a_n = a_1 + (n - 1)\,d$
$a_6 = a_1 + (6 - 1)\,d$
$15 = 5 + (6 - 1)\,d$
$d = 2$
$a_2 = 7$, $a_3 = 9$, $a_4 = 11$, $a_5 = 13$

These numbers are called **arithmetic means** between a_1 and a_6 .

Arithmetic Series

The sum of the first n terms:

$$S_n = \sum_{i=1}^{n} a_i = \frac{n}{2}(a_1 + a_n)$$

e.g. $1 + 2 + 3 + 4 + \bullet \bullet \bullet + 100$
$$= \frac{100}{2}(1 + 100) = 5050$$

(4). Geometric Sequences

$a_n = a_1 \bullet r^{n-1}$ r is the common ratio

or $a_n = a_{n-1} \bullet r$ recursive definition

e.g. { 1, $\frac{1}{2}$, $\frac{1}{4}$, $\frac{1}{8}$, $\frac{1}{16}$ }

e.g. { 2, -4, 8, -16, 32, $\bullet \bullet \bullet$ }

e.g. Find the **geometric means** between $a_1 = 5$ and $a_5 = 80$

$a_n = a_1 \bullet r^{n-1}$
$a_5 = a_1 \bullet r^{5-1}$
$80 = 5 \bullet r^4$
$r^4 = 16$
$r = 2$ or $r = -2$
$a_2 = 10$, $a_3 = 20$, $a_4 = 40$
or $a_2 = -10$, $a_3 = 20$, $a_4 = -40$

Geometric Series

Since $\dfrac{(1 - r^n)}{1 - r} = 1 + r + r^2 + \bullet \bullet \bullet + r^{n-1}$

The sum of the first n terms:

$$S_n = \sum_{i=1}^{n} a_i = a_1 + a_1r + a_1r^2 + \bullet \bullet \bullet + a_1r^{n-1} = \frac{a_1(1 - r^n)}{1 - r}$$

e.g. $5 + \dfrac{5}{2} + \dfrac{5}{4} + \dfrac{5}{8} + \dfrac{5}{16}$

$a_1 = 5$, $r = \dfrac{1}{2}$, $n = 5$

$$S_n = \frac{a_1(1 - r^n)}{1 - r} = \frac{5(1 - (\frac{1}{2})^5)}{(1 - \frac{1}{2})} = \frac{5 \bullet \frac{31}{32}}{\frac{1}{2}} = \frac{155}{16}$$

(5). Infinite Series

1. An infinite arithmetic series has no limit.
2. An infinite geometric series has no limit when $|r| \geq 1$.
3. An infinite geometric series has a limit when $|r| < 1$.

$$S_n = \sum_{i=1}^{\infty} a_i = \frac{a_1}{1 - r}$$

e.g. Find the limit of $0.252525 \bullet \bullet \bullet$
$0.252525 \bullet \bullet \bullet = 0.25 + 0.0025 + 0.000025 + \bullet \bullet \bullet$

$a_1 = 0.25$, $r = \dfrac{1}{100} = 0.01$

$$S_n = \frac{a_1}{1 - r} = \frac{0.25}{1 - 0.01} = \frac{0.25}{0.99} = \frac{25}{99}$$

4. The Number e

$$\sum_{n=0}^{\infty} \frac{1}{n!} = 1 + \frac{1}{1!} + \frac{1}{2!} + \frac{1}{3!} + \bullet \bullet \bullet = e$$

$e = 2.718281828 \bullet \bullet \bullet$

II. RELATIONS AND FUNCTIONS

1. Functions

Relation (Not a Function)

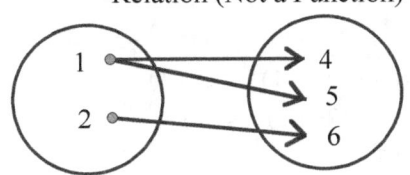

Function

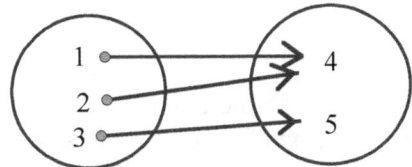

One-to-One Function

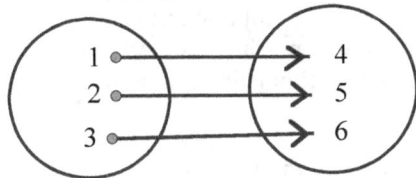

Domain and Range

e.g. $y = x^2$ Domain: $\{x \mid x \text{ all real numbers}\}$
                  Range: $\{y \mid y \geq 0\}$

e.g. $y = \sqrt{x}$ Domain: $\{x \mid x \geq 0\}$
                  Range: $\{y \mid y \geq 0\}$

e.g. $y = \dfrac{1}{x^2 - 9}$ Domain: $\{x \mid x \text{ all reals except } \pm 3\}$

e.g. $y = \dfrac{1}{\sqrt{x - 3}}$ Domain: $\{x \mid x > 3\}$

Interval Notation

e.g. $(2, 5)$ represents $\{ x \mid 2 < x < 5 \}$
      $[2, 5]$ represents $\{ x \mid 2 \leq x \leq 5 \}$
      $(2, 5]$ represents $\{ x \mid 2 < x \leq 5 \}$
      $[2, 5)$ represents $\{ x \mid 2 \leq x < 5 \}$

e.g. $(- \infty , \infty)$ represents $\{ x \mid x \text{ all real numbers} \}$
      $(- \infty , - 5)$ represents $\{ x \mid x < - 5 \}$
      $[5 , \infty)$ represents $\{ x \mid x \geq 5 \}$
      $(- \infty , - 5) \cup [5 , \infty)$ represents $\{ x < -5 \text{ or } x \geq 5 \}$

Function Notation

e.g.   If $f(x) = \dfrac{2x}{x - 1}$

then $f(5) = \dfrac{2 \bullet 5}{5 - 1} = \dfrac{10}{4} = \dfrac{5}{2}$

$f(a + 2) = \dfrac{2(a + 2)}{(a + 2) - 1} = \dfrac{2a + 4}{a + 1}$

$f(x^2) = \dfrac{2(x^2)}{(x^2) - 1} = \dfrac{2x^2}{x^2 - 1}$

Vertical Line Test:
If any vertical line intersects the graph at only one point, then the relation is a function.

e.g. $y^2 = x$ is equivalent to $y = \pm\sqrt{x}$. It is not a function.

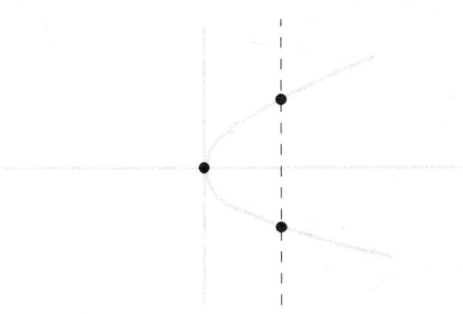

Horizontal Line Test:
If any horizontal line intersects the graph at only one point, then the function is a **one-to-one function**.

e.g. $y = x^2$ is a function, but not a one-to-one function.

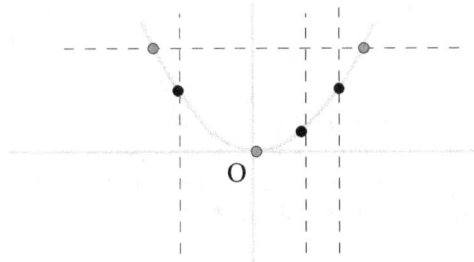

If we restrict the domain of $y = x^2$ to $x \geq 0$, then the function is a one-to-one funcion.

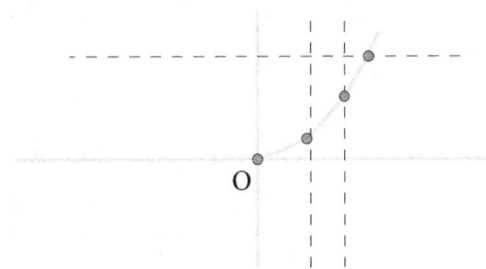

Onto

A function from set A to set B is onto if the range of the function is equal to set B.

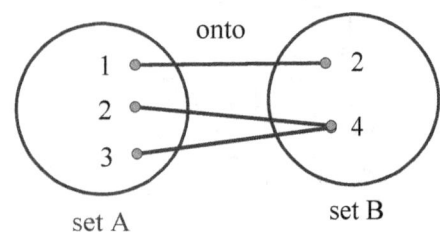

e.g. Set A and set B both are all real numbers.
    $y = x^3$ is onto and one-to-one.
    $y = x^2$ is neither onto nor one-to-one.
(the range $y \geq 0$ is not equal to all real numbers)
(not passing the horizontal line test).

e.g. Set A and set B both are all real numbers.
    $y = \sqrt{x}$ is one-to-one function but not onto.

2. Composition of Functions

$$(f \circ g)(x) = f(g(x))$$
e.g. $f(x) = x^2 - 1$, $g(x) = x + 1$
  $(f \circ g)(x) = f(g(x)) = f(x + 1) = (x + 1)^2 - 1 = x^2 + 2x$
  $(f \circ g)(2) = f(g(2)) = f(2 + 1) = f(3) = 3^2 - 1 = 8$
  $(g \circ f)(x) = g(f(x)) = g(x^2 - 1) = (x^2 - 1) + 1 = x^2$
  $(g \circ f)(2) = g(f(2)) = g(2^2 - 1) = g(3) = 3 + 1 = 4$
Note: $(f \circ g)(x) \neq (g \circ f)(x)$

3. Inverse Functions

For every one-to-one function f(x), there is an inverse function $f^{-1}(x)$. (passing both vertical and horizontal line tests)

e.g. $y = x^2$
It is a function. (passing the vertical line test)
But it is not a one-to-one function. (fails the horizonal line test)
The domain of the inverse function is the range of the original function.

e.g.   Original $f(x) = \{(1, 1), (2, 4), (3, 9)\}$
          Domain: $\{x \mid x = 1, 2, 3\}$
          Range: $\{y \mid y = 1, 4, 9\}$
          Inverse $f^{-1}(x) = \{(1, 1), (4, 2), (9, 3)\}$
          Domain: $\{x \mid x = 1, 4, 9\}$
          Range: $\{y \mid y = 1, 2, 3\}$

e.g. $f(x) = 3x + 5$ find $f^{-1}(x)$
     $y = 3x + 5$ express $f^{-1}(x)$ as y
     $x = 3y + 5$ interchange x and y
     $y = \dfrac{x - 5}{3}$ solve for y in terms of x

     $f^{-1}(x) = \dfrac{x - 5}{3}$

The graph of $f^{-1}(x)$ is the reflection of f(x) in the y = x.

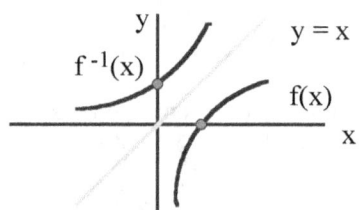

4. Functions under a Transformation

Translation:
$y = f(x)$ _____Ta,b_____ $y = f(x - a) + b$

Reflection:
$y = f(x)$ ___rx-axis___ $y = - f(x)$

$y = f(x)$ ___ry-axis___ $y = f(- x)$

Dilation:
$y = f(x)$ ___vertical stretch if $a > 1$___ $y = af(x)$
vertical shrink if $0 < a < 1$

The transformation rules for functions are different
from the transformation rules for images.
Tips: Use the graphing calculator to verify the answer.

e.g.

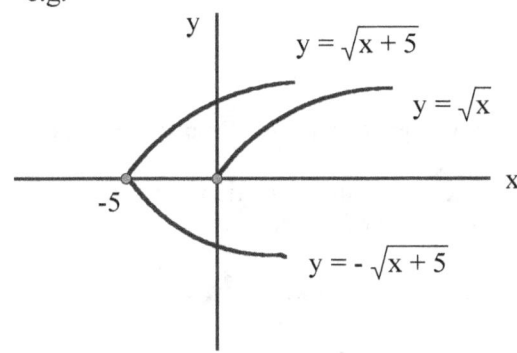

$y = \sqrt{x}$ ___moved 5 units to the left___ $y = \sqrt{x + 5}$

$y = \sqrt{x + 5}$ ___reflected in the x-axis___ $y = - \sqrt{x + 5}$

e.g.

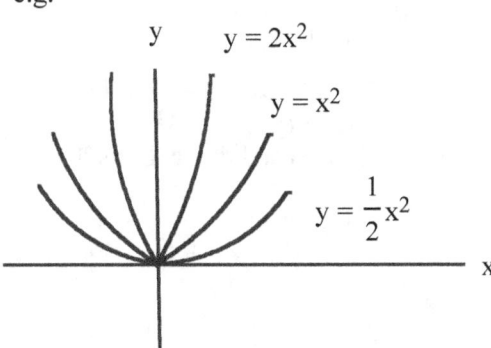

$y = x^2$ ___stretched vertically by a factor of 2___ $y = 2x^2$

$y = x^2$ ___shrunk vertically by a factor of 1/2___ $y = \frac{1}{2}x^2$

5. Important Functions and Relations

(1). Direct Variation

A straight line passing through the Origin

$$y = mx \qquad \text{or} \qquad \frac{y}{x} = m$$

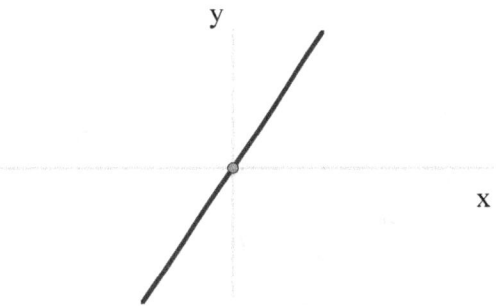

To solve a problem, use $\dfrac{x_1}{x_2} = \dfrac{y_1}{y_2}$

e.g. The distance varies directly as the time when a car
travels at a constant speed.
$$d = s \bullet t \qquad (s \text{ is a constant})$$

(2). Inverse Variations: (Function)

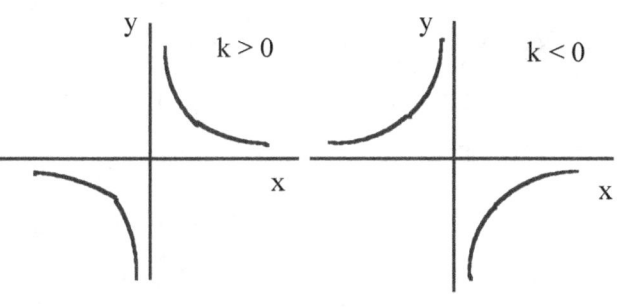

$$xy = k \quad \text{or} \quad y = \frac{k}{x}$$

To solve a problem, use $x_1 \bullet y_1 = x_2 \bullet y_2$

e.g. The speed varies inversely to the time when a car
travels over a certain distance.
$$s \bullet t = d \qquad (d \text{ is a constant})$$

(3). Absolute Value Functions

$$y = |x|$$

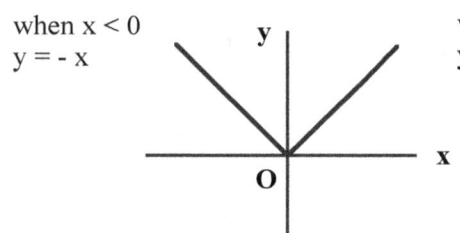

when $x < 0$
$y = -x$

when $x \geq 0$,
$y = x$

e.g. $y = |2x - 4|$

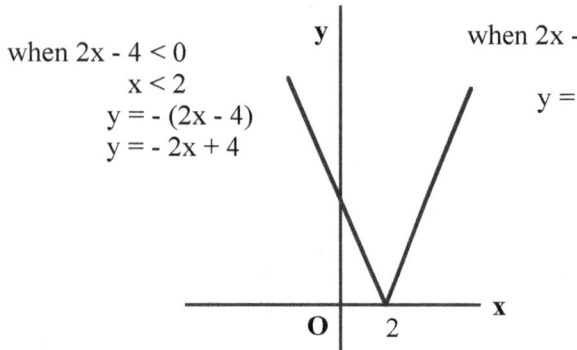

when $2x - 4 < 0$
$\qquad x < 2$
$\qquad y = -(2x - 4)$
$\qquad y = -2x + 4$

when $2x - 4 \geq 0$
$\qquad x \geq 2$
$\qquad y = 2x - 4$

(4). Quadratic Functions and Parabolas

General Form:

$$y = f(x) = ax^2 + bx + c \qquad \text{where } a \neq 0$$

(1). Axis of Symmetry: $x = -\dfrac{b}{2a}$

(2). Vertex (Turning Point): (x, y)

$$x = -\frac{b}{2a} \quad , \quad y = f(x) = f\left(-\frac{b}{2a}\right)$$

(3). Opening:

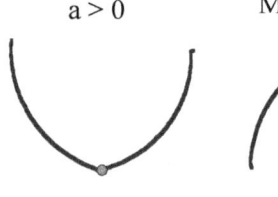

$a > 0$

Minimum point

Maximum point

$a < 0$

e.g. $y = 12x - 2x^2$
$\qquad y = -2x^2 + 12x \qquad$ write in standard form
$\qquad a = -2, \ b = 12, \ c = 0$

(1). Axis of Symmetry: $x = -\dfrac{b}{2a} = -\dfrac{12}{2(-2)} = 3$

(2). Vertex: $x = 3, \ y = -2(3)^2 + 12(3) = 18$
$\qquad\qquad (3, 18)$

(3). Opening: $a = -2 < 0$
$\qquad$ It has a maximum of 18 at $x = 3$.

Solve Quadratic Inequalities in Two Variables

e.g. $x^2 - 2x < 5 + y$
$\qquad$ Rewrite it in the standard form:
$\qquad\quad y > x^2 - 2x - 5$
$\qquad$ Graph $y = x^2 - 2x - 5$ in dashed line.
$\qquad$ The shaded region above the curve is the solution

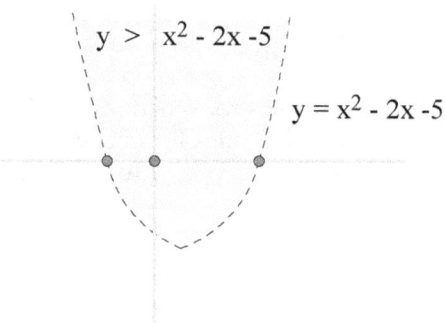

$y > x^2 - 2x - 5$

$y = x^2 - 2x - 5$

(5). Equations of Circles (Relation, not Function)

The center-radius equation of a circle
with radius r and center (h, k)

$$(x - h)^2 + (y - k)^2 = r^2$$

e.g. $x^2 + y^2 + 4x - 6y - 12 = 0$
$\qquad$ Find its center and radius, and graph it.
$\qquad x^2 + 4x + y^2 - 6y = 12$
$\qquad$ Complete the squares

$$x^2 + 4x + \left(\frac{4}{2}\right)^2 + y^2 - 6y + \left(\frac{-6}{2}\right)^2$$

$$= 12 + \left(\frac{4}{2}\right)^2 + \left(\frac{-6}{2}\right)^2$$

$$x^2 + 4x + 4 + y^2 - 6y + 9 = 12 + 4 + 9$$

$$(x + 2)^2 + (y - 3)^2 = 5^2$$

center is $(-2, 3)$ and the radius is 5

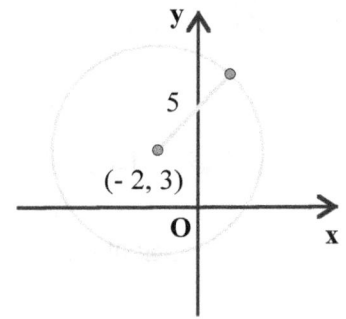

(6). Exponential Functions and Equations

$$y = a^x \qquad \text{where } a > 0 \text{ and } a \neq 1$$

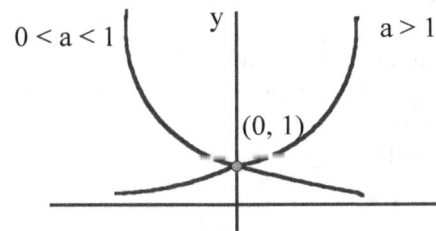

(1). Domain: { x | x all real numbers }
   Range: { y | y > 0 }
(2). a > 1 , the function is increasing (exponential growth);
   a < 1 , the function is decreasing (exponential decay).
(3). when x = 0, y = 1   (the graphs of the
       exponential functions pass point (0, 1)).
(4). x-axis is the horizontal asymptote.
(5). $y = (\frac{1}{a})^x$ and $y = a^x$ are symmetric in the y-axis.

General Form:

$$y = k \cdot a^x \qquad \text{where } a > 0 \text{ and } a \neq 1 , k \text{ is a constant}$$

Exponential Models:

e.g. $A = A_0 e^{-0.025t}$

   A_0 is the original amount;
   A is the amount at time t;
Find the half-life of this exponential decay.

$$\text{half-life: } A = \frac{1}{2}A_0 = 0.5A_0$$

$$0.5A_0 = A_0 e^{-0.025t}$$

$$0.5 = e^{-0.025t}$$

$$\ln 0.5 = -0.025t \ln e \qquad (\ln e = 1)$$

$$t = \frac{\ln 0.5}{-0.025} = 27.726$$

Exponential Equations

e.g. $9^{x+1} = 27^x$

   $3^{2(x+1)} = 3^{3x}$ transform to same base

   $2(x+1) = 3x$

   $x = 2$ check: true

e.g. $3^x = 5$

   $\log 3^x = \log 5$

   $x \log 3 = \log 5$

   $x = \frac{\log 5}{\log 3} = 1.465$

(7). Logarithmic Functions and Equations

$$y = \log_a x \text{ is equivalent to } x = a^y$$
e.g. $2 = \log_5 x$ is equivalent to $x = 5^2 = 25$

$y = \log_a x$ is the inverse function of $y = a^x$

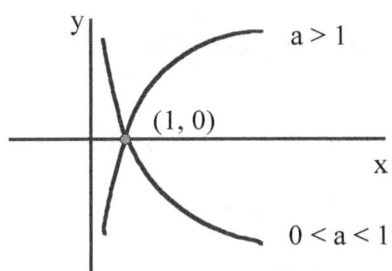

Domain: { x | x > 0 }
Range: { y | y all real numbers }

$$\log_a 1 = 0 , \quad \log_a a = 1$$
$$\log_a AB = \log_a A + \log_a B$$
$$\log_a \frac{A}{B} = \log_a A - \log_a B$$

$$\log_a A^n = n \cdot \log_a A , \quad \log_a \sqrt[n]{A} = \frac{1}{n}\log_a A$$

e.g. $\log_2 \sqrt[3]{2} = \log_2 2^{\frac{1}{3}} = \frac{1}{3}\log_2 2 = \frac{1}{3}$

e.g.   If $\log_5 x = 2$, what is the value of $\sqrt{x}$?
   $\log_5 x = 2$ is equivalent to $x = 5^2 = 25$
   $\sqrt{x} = \sqrt{25} = 5$

Common Logarithms

$$\log A = \log_{10} A$$

e.g. $\log 100 = \log 10^2 = 2$
e.g. $\log 0.01 = \log 10^{-2} = -2$
e.g. $\log 123 = \log 1.23 \cdot 10^2 = 2 + \log 1.23$
e.g.   If $\log A = 2$ and $\log B = 3$, then

$$\log \frac{\sqrt{A}}{B^3} = \frac{1}{2}\log A - 3\log B = \frac{1}{2} \cdot 2 - 3 \cdot 3 = -8$$

e.g. $\log_2 5 = \frac{\log 5}{\log 2} \approx 2.32$

Natural Logarithms

$$\ln A = \log_e A$$

$$e = \sum_{n=0}^{\infty} \frac{1}{n!} = 1 + \frac{1}{1!} + \frac{1}{2!} + \frac{1}{3!} + \cdots$$

$$e = 2.718281828\cdots$$

$$\ln 10 \approx 2.3 \qquad \log e \approx 0.4343$$

e.g. $\ln x = 5$
$x = e^5 \approx 148.4$

e.g. $e^{2\ln 5} = e^{\ln 5^2} = 5^2 = 25$

Change of Base Formula

$$\log_a A = \frac{\log A}{\log a}$$

or $$\log_a A = \frac{\ln A}{\ln a}$$

e.g. $$\log_2 5 = \frac{\log 5}{\log 2} \approx 2.3219$$

Logarithmic Equations

e.g. $\log_x 4 + \log_x 9 = 2$
$\log_x 4\cdot 9 = 2$
$x^2 = 36$
$x = 6 \qquad (\; x = -6 \text{ rejected }\;)$

e.g. $\log_4(x^2 + 3x) - \log_4(x + 5) = 1$

$$\log_4 \frac{x^2 + 3x}{x + 5} = \log_4 4$$

$$\frac{x^2 + 3x}{x + 5} = 4$$

$x^2 + 3x = 4x + 20$
$x^2 - x - 20 = 0$
$(x + 4)(x - 5) = 0$
$x = -4 \; \text{ or } \; x = 5 \qquad \text{check: true}$

PART 2. Trigonometry

III. TRIGONOMETRIC FUNCTIONS

1. Degrees and Radians

The degree measure of a circle is 360.
The degree measure of a semicircle is 180.

$$1° = 60' \quad \text{(minutes)}$$

e.g. $0.3° = 0.3\cdot 60 = 18'$

$$45' = \frac{45}{60} = 0.75°$$

Radian is a different unit to measure the angle and the arc.
The radian measure of a circle is 2π.
The radian measure of a semicircle is π.

2π (radians) = $360°$
π (radians) = $180°$

$$1 \text{ (radian)} = \frac{180°}{\pi} \approx 57.3°$$

Conversion:

use $\dfrac{\pi}{180°} = 1$ or $\dfrac{180°}{\pi} = 1$

e.g. $60° = 60°\cdot \dfrac{\pi}{180°} = \dfrac{\pi}{3}$

e.g. $\dfrac{\pi}{4} = \dfrac{\pi}{4}\cdot \dfrac{180°}{\pi} = 45°$

Coterminal Angles:

Coterminal angles have the same terminal side.
They differ $360°$ or a multiple of $360°$.

e.g. $30°$, $390°$, and $750°$ are coterminal angles.

e.g. $-60°$ and $300°$ are coterminal angles.

Arc Length:

$s = r \cdot \theta$ \qquad where r is radius and θ in radian

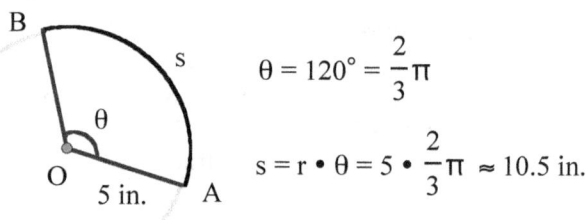

$$\theta = 120° = \frac{2}{3}\pi$$

$$s = r \cdot \theta = 5 \cdot \frac{2}{3}\pi \approx 10.5 \text{ in.}$$

These problems can also be solved by
using fractions or ratios to the circle.

2. Trigonometric Fuctions

Trigonometric Ratios and Basic Functions

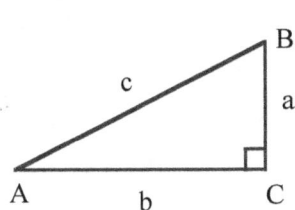

$$\sin A = \frac{\text{Opp}}{\text{Hyp}} = \frac{a}{c}$$

$$\cos A = \frac{\text{Adj}}{\text{Hyp}} = \frac{b}{c}$$

$$\tan A = \frac{\text{Opp}}{\text{Adj}} = \frac{a}{b}$$

Reciprocal Functions:

$$\cot A = \frac{1}{\tan A} \ , \quad \sec A = \frac{1}{\cos A} \ , \quad \csc A = \frac{1}{\sin A}$$

Exact Values to Remember:

θ (degree)	0°	30°	45°	60°	90°
θ	0	$\pi/6$	$\pi/4$	$\pi/3$	$\pi/2$
$\sin\theta$	0	1/2	$\sqrt{2}/2$	$\sqrt{3}/2$	1
$\cos\theta$	1	$\sqrt{3}/2$	$\sqrt{2}/2$	1/2	0
$\tan\theta$	0	$\sqrt{3}/3$	1	$\sqrt{3}$	undefined

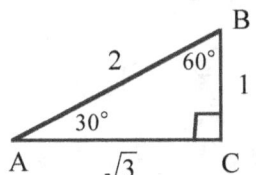
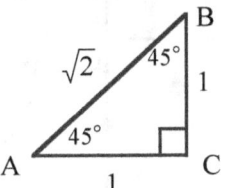

Pythagorean Triples:

3, 4, 5 and 5, 12, 13

e.g. If $\sin\theta = \frac{3}{5}$, then $\cos\theta = \frac{4}{5}$, $\tan\theta = \frac{3}{4}$

If $\tan\theta = \frac{5}{12}$, then $\sin\theta = \frac{5}{13}$, $\cos\theta = \frac{12}{13}$

Cofunctions

$\sin\theta = \cos(90° - \theta)$,
$\tan\theta = \cot(90° - \theta)$
$\sec\theta = \csc(90° - \theta)$

e.g. $\sin 30° = \cos 60°$
e.g. If $\sin A = \cos B$, then $A + B = 90°$
e.g. If $\sin 2A = \cos 4A$, find A
 $2A + 4A = 90°$ $A = 15°$

Unit Circle

Center at the origin ; Radius of one unit.

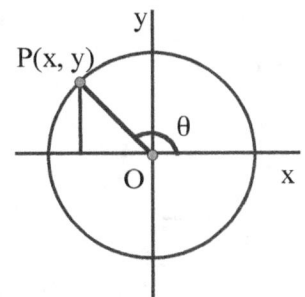
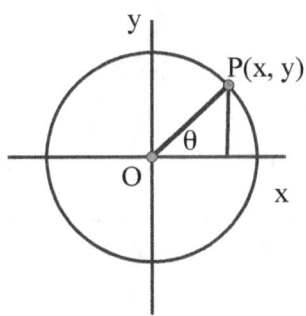

For Angles of Any Degree:

$$\cos\theta = x \ , \quad \sin\theta = y \ , \quad \tan\theta = \frac{y}{x}$$

In general, a point is not on a Unit Circle:

$$\cos\theta = \frac{x}{\sqrt{x^2 + y^2}} \ , \quad \sin\theta = \frac{y}{\sqrt{x^2 + y^2}} \ , \quad \tan\theta = \frac{y}{x}$$

Reference Angle

An acute angle formed by the terminal side of the given angle and the x-axis.

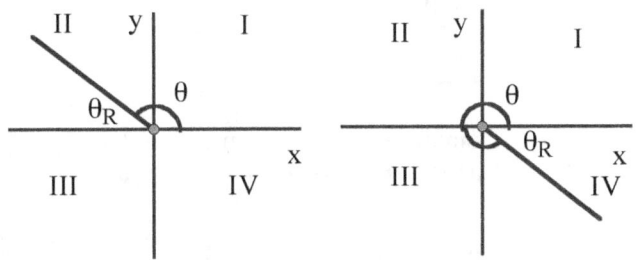

Quadrant	I	II	III	IV
θ_R	θ	$180° - \theta$	$\theta - 180°$	$360° - \theta$
$\sin\theta$	$\sin\theta_R$	$\sin\theta_R$	$- \sin\theta_R$	$- \sin\theta_R$
$\cos\theta$	$\cos\theta_R$	$- \cos\theta_R$	$- \cos\theta_R$	$\cos\theta_R$
$\tan\theta$	$\tan\theta_R$	$- \tan\theta_R$	$\tan\theta_R$	$- \tan\theta_R$

e.g. $\cos 120° = - \cos 60° = -\frac{1}{2}$ ($\theta_R = 180° - 120° = 60°$)

$\cos 240° = - \cos 60° = -\frac{1}{2}$ ($\theta_R = 240° - 180° = 60°$)

$\cos 300° = \cos 60° = \frac{1}{2}$ ($\theta_R = 360° - 300° = 60°$)

IV. TRIGONOMETRIC GRAPHS

1. Graphs of Trigonometric Functions

$y = \sin x$

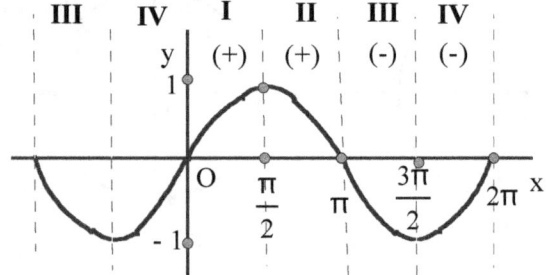

Domain: { x | x all real numbers }
Range: { y | - 1 ≤ y ≤ 1 }

$y = \cos x$

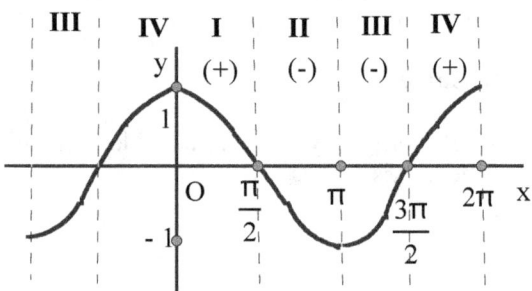

Domain: { x | x all real numbers }
Range: { y | - 1 ≤ y ≤ 1 }

$y = \tan x$

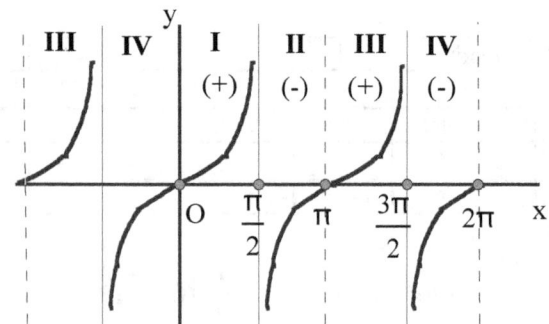

Domain: $\{ x \mid x \neq \dfrac{\pi}{2} + n\pi \ \text{for n an integer} \}$

Range: { y | y all real numbers }

2. Graphs of the Reciprocal Functions

when $f(x) \rightarrow 0$, $\dfrac{1}{f(x)} \rightarrow \infty$

$0 < f(x) < 1$, $\dfrac{1}{f(x)} > 1$

$f(x) = 1$, $\dfrac{1}{f(x)} = 1$

$f(x) > 1$, $\dfrac{1}{f(x)} < 1$

$f(x) \rightarrow \infty$, $\dfrac{1}{f(x)} \rightarrow 0$, etc.

e.g. $y = \csc x = \dfrac{1}{\sin x}$

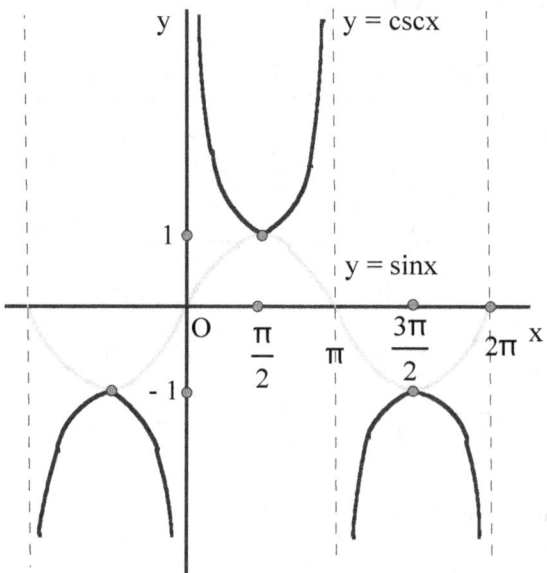

$y = \csc x$

Domain: { x | x : all real numbers except nπ }
Range: { y | y ≤ - 1 or y ≥ 1 }

Use graphing calculator to see y = cotx and
y = secx.

3. Amplitude, Period, and Frequency:

y = asinbx   and   **y = acosbx**

amplitude $= |a|$, frequency $= |b|$, period $= \dfrac{2\pi}{|b|}$

e.g. $y = \cos x$
amplitude $= 1$, frequency $= 1$, period $= 2\pi$

e.g. $y = -3\sin 2x$

amplitude $= 3$, frequency $= 2$, period $= \dfrac{2\pi}{2} = \pi$

y = tanbx

frequency $= |b|$, period $= \dfrac{\pi}{|b|}$

e.g. $y = \tan x$ period $= \pi$

e.g. $y = \tan 2x$ period $= \dfrac{\pi}{2}$

The graph of y = asinb(x + c) + d

e.g. Graph $y = 5\sin 2(x + \dfrac{\pi}{4}) + 3$

Amplitude $= 5$, Frequency $= 2$, Period $= \pi$

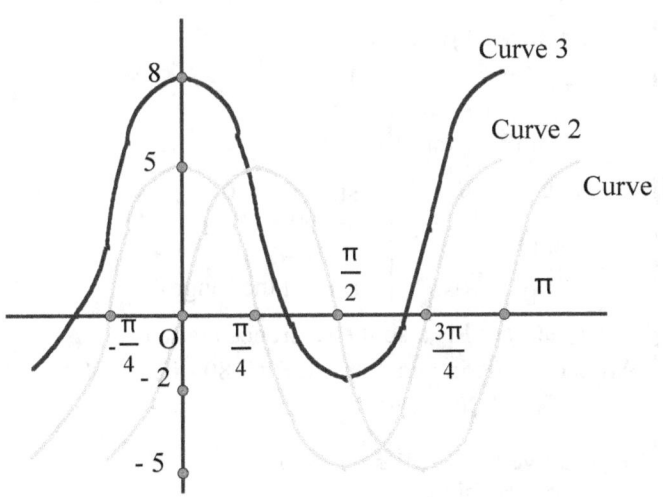

Curve 1: $y = 5\sin 2x$

Curve 2: $y = 5\sin 2(x + \dfrac{\pi}{4})$ shift $\dfrac{\pi}{4}$ to the left (phase shift $-\dfrac{\pi}{4}$)

Curve 3: $y = 5\sin 2(x + \dfrac{\pi}{4}) + 3$ shift 3 units up

4. Inverse Trigonometric Functions

$y = \arcsin x$ or $y = \sin^{-1}x$

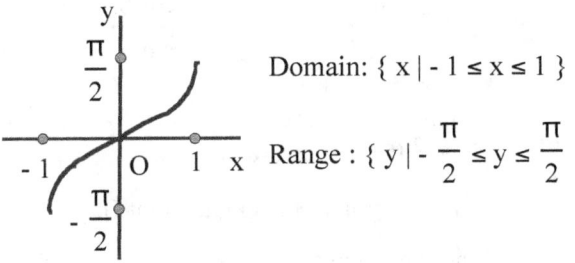

Domain: $\{\, x \mid -1 \le x \le 1 \,\}$

Range : $\{\, y \mid -\dfrac{\pi}{2} \le y \le \dfrac{\pi}{2} \,\}$

$y = \arccos x$ or $y = \cos^{-1}x$

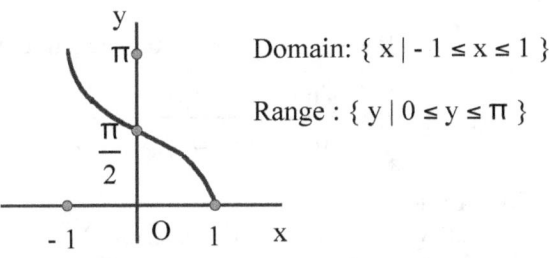

Domain: $\{\, x \mid -1 \le x \le 1 \,\}$

Range : $\{\, y \mid 0 \le y \le \pi \,\}$

$y = \arctan x$ or $y = \tan^{-1}x$

Domain: $\{\, x \mid x \text{ all real numbers} \,\}$

Range: $\{\, y \mid -\dfrac{\pi}{2} < y < \dfrac{\pi}{2} \,\}$

e.g. Find $\sin(\arccos \dfrac{1}{2})$

Let $\theta = \arccos \dfrac{1}{2}$ $(0 \le \theta \le \pi)$

$\cos \theta = \dfrac{1}{2}$

since $0 \le \theta \le \pi$, $\theta = 60°$

$\sin \theta = \dfrac{\sqrt{3}}{2}$

V. TRIGONOMETRIC APPLICATIONS

1. Trigonometric Identities

$$\tan\theta = \frac{\sin\theta}{\cos\theta} \quad , \quad \cot\theta = \frac{\cos\theta}{\sin\theta}$$

$$\sin^2\theta + \cos^2\theta = 1$$
$$\tan^2\theta + 1 = \sec^2\theta$$
$$\cot^2\theta + 1 = \csc^2\theta$$

e.g. $\sin\theta = \frac{5}{13}$ and θ is in Quadrant II.

Find the value of (1) $\cos\theta$ and (2) $\tan\theta$.

(1) $(\frac{5}{13})^2 + \cos^2\theta = 1$

$\cos\theta = -\frac{12}{13}$ ($\cos\theta$ is negative in Q II)

(2) $\tan\theta = \frac{\sin\theta}{\cos\theta} = \frac{5/13}{-12/13} = -\frac{5}{12}$

Two Angles
$$\sin(A + B) = \sin A\cos B + \cos A\sin B$$
$$\sin(A - B) = \sin A\cos B - \cos A\sin B$$
$$\cos(A + B) = \cos A\cos B - \sin A\sin B$$
$$\cos(A - B) = \cos A\cos B + \sin A\sin B$$

e.g. The following identities are true for all values of θ:
$\sin(-\theta) = -\sin\theta, \ \cos(-\theta) = \cos\theta, \ \tan(-\theta) = -\tan\theta$
$\sin(90° - \theta) = \cos\theta, \quad \cos(90° - \theta) = \sin\theta$

Double Angles
$$\sin 2A = 2\sin A\cos A$$
$$\cos 2A = \cos^2 A - \sin^2 A$$
$$= 2\cos^2 A - 1$$
$$= 1 - 2\sin^2 A$$
$$\tan 2A = \frac{2\tan A}{1 - \tan^2 A}$$

Half Angles
$$\sin\frac{1}{2}A = \pm\sqrt{\frac{1 - \cos A}{2}}$$
$$\cos\frac{1}{2}A = \pm\sqrt{\frac{1 + \cos A}{2}}$$
$$\tan\frac{1}{2}A = \pm\sqrt{\frac{1 - \cos A}{1 + \cos A}}$$

Find θ in different Quadrants in terms of θ_R

Quadrant	I	II	III	IV
θ	θ_R	$180° - \theta_R$	$180° + \theta_R$	$360° - \theta_R$

e.g. Find the exact value of $\cos 105°$
$$\cos 105° = \cos(60° + 45°)$$
$$= \cos 60°\cos 45° - \sin 60°\sin 45°$$
$$= \frac{1}{2}\cdot\frac{\sqrt{2}}{2} - \frac{\sqrt{3}}{2}\cdot\frac{\sqrt{2}}{2}$$
$$= \frac{\sqrt{2} - \sqrt{6}}{4}$$

e.g. $\sin x = \frac{4}{5}$, x is an acute angle.

Find the value of $\cos 2x$ and $\sin 2x$.
(1) $\cos 2x = 1 - 2\sin^2 x$
$$= 1 - 2(\frac{4}{5})^2 = -\frac{7}{25}$$

(2) $\cos x = \sqrt{1 - \sin^2 x} = \frac{3}{5}$ (acute angle)

$\sin 2x = 2\sin x\cos x$
$$= 2(\frac{4}{5})(\frac{3}{5}) = \frac{24}{25}$$

or $\sin 2x = \sqrt{1 - \cos^2 2x} = \frac{24}{25}$

(sine is positive when 2x is in either Quadrant I or II,)

2. Trigonometric Equations

e.g. Solve for θ , $0° \le \theta < 180°$
$$2\cos^2\theta - \sin\theta = 1$$
$$2(1 - \sin^2\theta) - \sin\theta = 1$$
$$2 - 2\sin^2\theta - \sin\theta = 1$$
$$2\sin^2\theta + \sin\theta - 1 = 0$$
$$(2\sin\theta - 1)(\sin\theta + 1) = 0$$

$2\sin\theta - 1 = 0$	$\sin\theta + 1 = 0$
$\sin\theta = \frac{1}{2}$	$\sin\theta = -1$
	$\theta = 270°$
$\theta = 30°$	(not in the range)

θ in Quadrant I is also the reference angle θ_R.
Another angle in Quadrant II, $\theta = 180 - \theta_R = 150°$
$$\{30°, 150°\}$$

e.g. Solve for θ , $0° \le \theta < 360°$
$$\cos 2\theta - \sin\theta = 1$$
$$1 - 2\sin^2\theta - \sin\theta = 1$$
$$\sin\theta(2\sin\theta + 1) = 0$$

$\sin\theta = 0$	$2\sin\theta + 1 = 0$
$\theta = 0°$ or $\theta = 180°$	$\sin\theta = -0.5$ no θ in Quadrant I
	Solve $\sin\theta_R = 0.5$, $\theta_R = 30°$
	$\theta = 180° + 30° = 210°$, or
	$\theta = 360° - 30° = 330°$

$$\{0°, 180°, 210°, 330°\}$$

3. Applications

For any triangle:

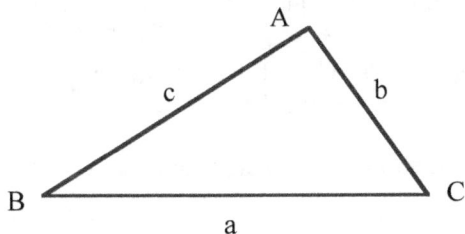

$$\textbf{Area} = \frac{1}{2}ab\sin C = \frac{1}{2}bc\sin A = \frac{1}{2}ca\sin B$$

Law of Sines:

$$\frac{a}{\sin A} = \frac{b}{\sin B} = \frac{c}{\sin C}$$

Law of Cosines:

$$c^2 = a^2 + b^2 - 2ab\cos C$$
$$\cos C = \frac{a^2 + b^2 - c^2}{2ab}$$

e.g. Find the area of $\triangle ABC$

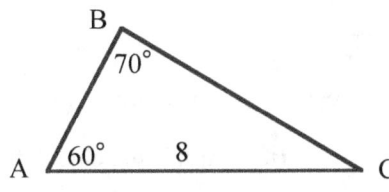

$$m\angle C = 180 - 70 - 60 = 50$$
$$\frac{AB}{\sin 50°} = \frac{AC}{\sin 70°}$$
$$AB = 6.52$$
$$\text{Area} = \frac{1}{2}bc\sin 60° = \frac{1}{2}\,8\,(6.5)\sin 60° = 22.6$$

e.g. **Ambiguous Case:**

In $\triangle ABC$, $m\angle A = 30$, $AB = 12$, $BC = 7$,
How many possible trianges can be constructed ?
$$\frac{BC}{\sin A} = \frac{AB}{\sin C}$$
$$\frac{7}{\sin 30°} = \frac{12}{\sin C}$$
$$\sin C = \frac{6}{7}$$
$$m\angle C = 59 \text{ or } m\angle C = 180 - 59 = 121$$
check: $m\angle A + m\angle C = 30 + 59 < 180$ OK
check: $m\angle A + m\angle C = 30 + 121 < 180$ OK
Two possible triangles can be constructed.

e.g. Find x.

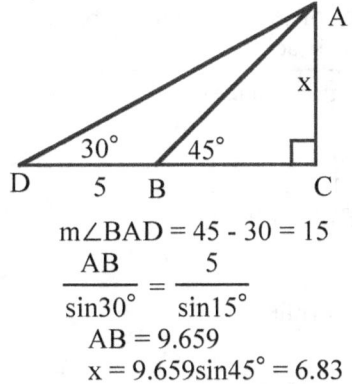

$$m\angle BAD = 45 - 30 = 15$$
$$\frac{AB}{\sin 30°} = \frac{5}{\sin 15°}$$
$$AB = 9.659$$
$$x = 9.659\sin 45° = 6.83$$

e.g. Two forces of 10 lb. and 15 lb. act on a body. Their resultant is 20 lb. Find the angle between the two applied forces.

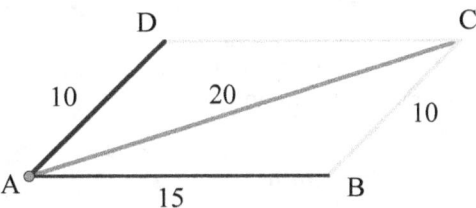

$BC = AD = 10$ (in a parallelogram opposite sides $\cong$)
In $\triangle ABC$, $\cos B = \dfrac{a^2 + c^2 - b^2}{2\,ac}$
$$= \frac{10^2 + 15^2 - 20^2}{2 \cdot 10 \cdot 15}$$
$$\cos B = -0.25$$
$$m\angle B = 104.5$$
$$m\angle BAD = 180 - 104.5 = 75.5$$
(Note: The angle between the two applied forces is $\angle BAD$, not $\angle B$.)

PART 3. Probability and Statistics

VI. PROBABILITY

1. Probability and Counting Principle

Sample Space: all possible outcomes
Event: the favorable outcomes

(1)The probability of a simple event

$$P(E) = \frac{\text{number of the outcomes of the event}}{\text{number of the outcomes of the sample space}}$$

$$P(E) = \frac{n(E)}{n(S)}$$

e.g. A bag contains 6 black balls and 4 white balls. What is the probability of selecting a black ball?

$$P(\text{Black}) = \frac{n(\text{Black})}{n(\text{Sample Space})} = \frac{6}{10}$$

Impossible Case: $P(E) = 0$
Certain Case: $P(E) = 1$
Negation: $P(\text{Not } E) = 1 - P(E)$

e.g. If $P(\text{rain}) = 30\%$,
then $P(\text{Not rain}) = 1 - P(\text{rain}) = 70\%$

(2) Counting Principle (2 or more activities)

If the first activity can occur in M ways and the second activity can occur in N ways, then both activities can occur in M•N ways.

e.g. 3 doors to a building, 2 stairways to the second floor. There are $3•2 = 6$ different ways to go.

Counting Principle for Probability

When A and B are indepenent events, the compound event of A and B has the probability
$$P(A , B) = P(A) • P(B)$$

e.g. 4 students. The probability of the tallest one in the first place (A) and the shortest one in the last place (B)

$$P(A , B) = P(A) • P(B) = \frac{1}{4} • \frac{1}{3} = \frac{1}{12}$$

2. Permutation and Combination

In a **permutation** the order of the objects is important.
(1) The permutaion of n objects taken n at a time
$$_nP_n = n! = n(n - 1)(n - 2) \ldots 2•1$$

e.g. Five letters A, B, C, D, E have 5! different arrangements. ($5! = 5•4•3•2•1 = 120$)

(2) The permutation of n objects taken n at a time
with r items identical : $\dfrac{n!}{r!}$

e.g. Five letters COLOR have $\dfrac{5!}{2!}$ different arrangements.

e.g. Seven letters FREEZER, three E's identical and two R's identical, have $\dfrac{7!}{3!•2!}$ different arrangements.

(3) The permutation of n objects taken r (r < n) at a time
$$_nP_r = n(n -1)(n -2)\ldots \quad (\text{r factors})$$

e.g. How many different arrangements of 1st, 2nd, and 3rd place are possible for 10 students?
$$_{10}P_3 = 10•9•8 = 720 \quad (\text{3 factors})$$

In a **combination** the order of the objects does not matter.

e.g. (A, B, C) and (C, B, A) are considered same.

(4) The combination of n objects taken r at a time
$$_nC_r = \frac{_nP_r}{r!} \quad (r \le n)$$
$$_nC_n = 1, \quad _nC_0 = 1, \quad _nC_1 = n, \quad _nC_r = {_nC_{n-r}}$$

e.g. $_{50}C_{48} = {_{50}C_2}$ (to simplify the calculation)

e.g. How many 3 player teams can be formed from 10 students?
$$_{10}C_3 = \frac{_{10}P_3}{3!} = \frac{10•9•8}{3•2•1} = 120$$

e.g. From 10 boys and 12 girls, how many different teams can be formed if 2 members must be boys and 3 members must be girls?
$$_{10}C_2 • {_{12}C_3} = \frac{10•9}{2•1} • \frac{12•11•10}{3•2•1} = 9900$$

Using Graphing Calculator

e.g. $5! = 120$
[5] [MATH] PRB / 4: ! [ENTER]

e.g. $_5P_3 = 60$
[5] [MATH] PRB / 2: $_nP_r$ [ENTER] 3 [ENTER]

e.g $_7C_4 = 35$
[7] [MATH] PRB / 3: $_nC_r$ [ENTER] 4 [ENTER]

3. Binomial Probability (Bernoulli Experiment)

If the probability of success is p and the probability of failure is $q = 1 - p$, then the probability of exactly r successes in n independent trials is
$$_nC_r \, p^r \, q^{n-r}$$

e.g. The probability of rain on any given day is 0.3.
(1) The probability of rain on exactly 2 of 7 days is
$P(2) = {_7C_2}(0.3)^2(0.7)^5$ $(q = 1 - 0.3 = 0.7)$
(2) The probability of rain at most 2 of 7 days is
$P(\text{at most } 2) = P(0) + P(1) + P(2)$
"at most 2 days" is same as "no more than 2 days"
(3) The probability of rain at least 4 of 7 days is
$P(\text{at least } 4) = P(4) + P(5) + P(6) + P(7)$
"at least 4 days" is same as "no less than 4 days"
(4) $P(0) + P(1) + P(2) + \bullet\ \bullet\ \bullet + P(6) + P(7) = 1$
e.g. $P(0) + P(1) + P(2) + \bullet\ \bullet\ \bullet + P(6) = 1 - P(7)$

Binomial Expansions:

$(x + y)^n = {_nC_0}x^ny^0 + {_nC_1}x^{n-1}y^1 + {_nC_2}x^{n-2}y^2 \bullet\ \bullet\ \bullet\ {_nC_n}x^0y^n$
There are $n + 1$ terms in the expansion.
The rth term is:
$${_nC_{(r-1)}}x^{n-(r-1)}y^{(r-1)}$$

e.g. $(x + y)^4 = 1\bullet x^4 + 4\bullet x^3\bullet y + 6\bullet x^2\bullet y^2 + 4\bullet x\bullet y^3 + 1\bullet y^4$

e.g. $(x + y)^8$
(1) the first term: $(n = 8, r = 1, r - 1 = 0)$
$_8C_0 x^{8-0}y^0 = 1\bullet x^8\bullet 1 = x^8$
(2) the last term: $(n = 8, r = 9, r - 1 = 8)$
$_8C_8 x^{8-8}y^8 = 1\bullet x^0\bullet y^8 = y^8$
(3) the middle term: $(n = 8, r = 5, r - 1 = 4)$
$_8C_4 x^{8-4}y^4 = 70x^4y^4$

e.g. $(2a - 1)^5$
The 3rd term is: $(n = 5, r = 3, r - 1 = 2)$
$_5C_2 (2a)^{5-2}(-1)^2 = 10\bullet(2a)^3\bullet 1 = 80a^3$

VII. STATISTICS

1. Statistics (Univariate Data)

Common Methods of Collecting Data:

Surveys: get information through questionnaires, interviews, etc.
Controlled Experiments: consist of two groups of data, one of them served as a benchmark.
Observations: watch and study on the phenomena, without influence on the responses.

Size of the Data:

A **population** consists of the set of all items of interest.
A **sample** is a subset of items chosen from a population. The sample must be large enough to be effective and must be chosen **randomly** to eliminate any **bias**.

Analyze Data:

First arrange the data in numerical order.
$$\text{Mean} = \text{Average} = \frac{\text{sum of the data values}}{\text{number of the data items}}$$
Median: the middle value when the data arranged in order
Mode: the value that appears most often

Percentile: a number that tells what percent of the total number of the data values are less than or equal to a given data point.
1st Quartile (25th percentile): the middle value of the lower half set of the data, aka. **Lower Quartile**
2nd Quartile (50th percentile): the **median**, aka. **Middle Quartile**
3rd Quartile (75th percentile): the middle value of the upper half set of the data, aka. **Upper Quartile**

Range: the difference between the highest value and the lowest value.
Interquartile Range: the difference between the 3rd quartile value and the 1st quartile value.

Outliers: Some data points far outside most of the points in the data set.
Outliers can strongly affect the mean value. When outliers exist, use median to represent the central tendency of the data.

e.g. Analyze the grades:

78, 85, 81, 95, 61, 85, 75, 88, 72, 100

First rearrange the data in numerical order:
61, 72, 75, 78, 81, 85, 85, 88, 95, 100
(make sure the number of items are same)

$$\text{Mean} = \frac{820}{10} = 82$$

$$\text{Median} = \frac{81 + 85}{2} = 83$$

(if the set has an even number of data values, take the average of the two middle values)

Mode = 85
Middle Quartile = Median = 83
Lower Quartile = 75
Upper Quartile = 88

Box-and-Whisker Plot :

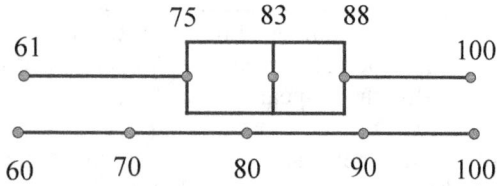

Range = 100 - 61, Interquartile Range = 88 - 75

Frequency Table :

Interval	Frequency
61 - 70	1
71 - 80	3
81- 90	4
91 - 100	2

Frequency Histogram :

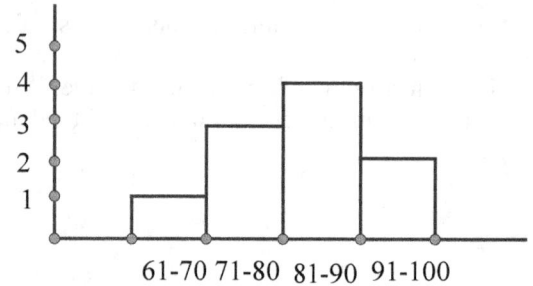

Variance for Populations:

$$v = \frac{1}{n} \sum_{i=1}^{n} (x_i - \bar{x})^2$$

Standard Deviation for Populations:
$$\delta = \sqrt{v} \qquad \text{(calculator symbol: } \delta_x)$$
n = number of the data items, $\bar{x}$ = mean, x_i = data values
Use graphing calculator to find $\bar{x}$ and δ_x.

Variance for Samples:

$$v = \frac{1}{n-1} \sum_{i=1}^{n} (x_i - \bar{x})^2$$

Standard Deviation for Samples:
$$S = \sqrt{v} \qquad \text{(calculator symbol: } S_x)$$
Use graphing calculator to find $\bar{x}$ and S_x.

The difference between δ_x and S_x is $\frac{1}{n}$ and $\frac{1}{n-1}$.
With same set of data, S_x is slightly greater than δ_x .
This difference is insignificant when the sample is large enough.

Normal Distribution:
The properties of a normal curve:
(1). Symmetric ;
(2). Mean, Median, and Mode have the same value.
(3). 68.2% of the data values between $\bar{x} - \delta$ and $\bar{x} + \delta$.
    95.4% of the data values between $\bar{x} - 2\delta$ and $\bar{x} + 2\delta$.
    99.8% of the data values between $\bar{x} - 3\delta$ and $\bar{x} + 3\delta$.

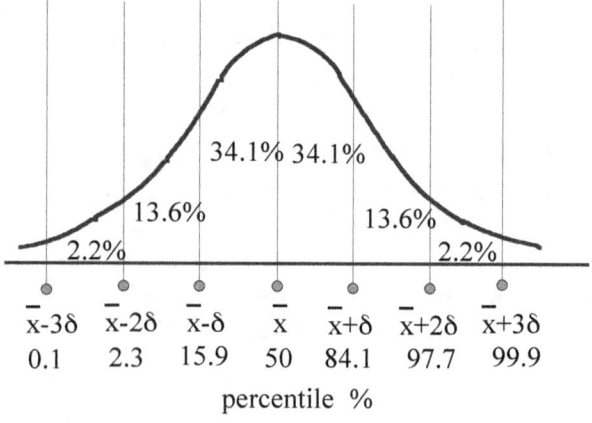

e.g. A normal distribution has a mean of 10.50 and a standard deviation of 0.75. What percent of the data values are in the range from 9.75 to 11.25 ?

$$9.75 = 10.50 - 0.75 = \bar{x} - \delta$$
$$11.25 = 10.50 + 0.75 = \bar{x} + \delta$$

There are 68.2% of the data values between 9.75 and 11.25.

e.g. The ages of the new teachers are normally distributed. 95.4% of the ages, centered about the mean, are between 24.6 and 37.4. Find the mean and the standard deviation.

$$\bar{x} = \frac{24.6 + 37.4}{2} = 31$$
$$(\bar{x} + 2\delta) - (\bar{x} - 2\delta) = 4\delta = 37.4 - 24.6 = 12.8$$
$$\delta = 3.2$$

Using Graphing Calculator

e.g. Find Mean and Standard Deviation
        95, 92, 86, 84, 78
(1) Clear List L_1
[STAT] EDIT / 4: ClrList [ENTER] [2nd] [L_1]
[ENTER]
(2) Enter data to L_1
[STAT] EDIT / 1: Edit ... [ENTER]
        enter the above data into list L_1
 (3) Display the One Variable Statistics
[STAT] CALC / 1: 1 - Var [ENTER] [2nd] [L_1]
[ENTER]
        $\bar{x} = 87$, $\delta x = 6$, $S_x = 6.7$
        $Q_1 = 81$, Med = 86 , $Q_3 = 93.5$

e.g. x_i values , f_i frequency

x_i	92	87	82	77	72	67	62
f_i	2	3	6	9	10	6	4

(1) clear List L_1 and List L_2
[STAT] EDIT / 4: ClrList [ENTER] [2nd] [L_1] [,]
[2nd] [L_2] [ENTER]
(2) Enter data to L_1 and L_2
[STAT] EDIT / 1: Edit ... [ENTER]
        enter the x_i data into list L_1 and the f_i data into L_2
(3) Display the **One Variable** Statistics
[STAT] CALC / 1: 1 - Var [ENTER] [2nd] [L_1] [,]
 [2nd] [L_2] [ENTER]
        $\bar{x} = 75$, $\delta x = 7.89$, $S_x = 7.99$,
        $Q_1 = 69.5$, Med = 74.5, $Q_3 = 82$
        $\bar{x} \neq$ Med   This is not a normal distribution.

Z-Score

$$\text{z-score} = \frac{x - \bar{x}}{\delta}$$

The z-score tells us how many standard deviations a data value x is above or below the mean.

Using Normal Curve to Estimate Binomial Probabilities

$$\bar{x} = np$$

$$\delta = \sqrt{npq} = \sqrt{np(1 - p)}$$

n: the number of trials , p: the probability of success
q: the probability of failure

Using Graphing Calculator

Computing the Normal Distribution Probability:
[2nd] [DISTR] 2: normalcdf (lowerbound, upperbound, x, δ) [ENTER]
This function computes the area under the normal curve from the lower bound to the upper bound.

e.g. Find the following probabilities when the probability of success $p = \dfrac{2}{5}$ and the number of trials n = 50 .

(1). P_1(at least 15 successes)

$$\bar{x} = np = 50 \cdot \frac{2}{5} = 20$$

$$\delta = \sqrt{np(1 - p)} = \sqrt{20 \cdot \frac{3}{5}} = 3.4641$$

normalcdf (14.5, 50.5, 20, 3.4641) = 0.944
Note: Use 14.5 instead of 15 for the lower bound and
        use 50.5 instead of 50 for the upper bound.

(2). P_2(fewer than 15 successes)
        normalcdf (0, 14.5, 20, 3.4641) = 0.056

(3). P_3(at most 15 successes)
        normalcdf (0, 15.5, 20, 3.4641) = 0.097
        Note: $P_1 + P_2 = 1$; $P_1 + P_3 \neq 1$

Algebra 2 and Trigonometry Review

2. Statistics (Bivariate Data)

Correlation: The relationship between two sets of data

Causation: The relationship in which one variable produces an effect on the other

Regression Modeling

Linear Regression: $y = ax + b$

Correlation Coefficient r :

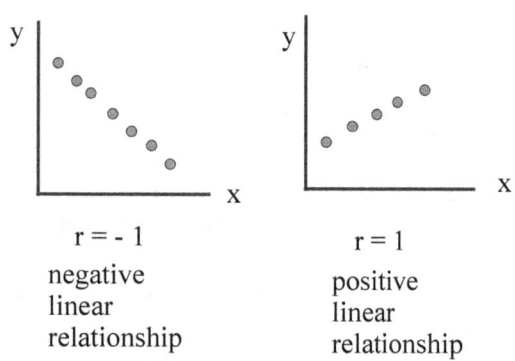

r = − 1

negative
linear
relationship

r = 1

positive
linear
relationship

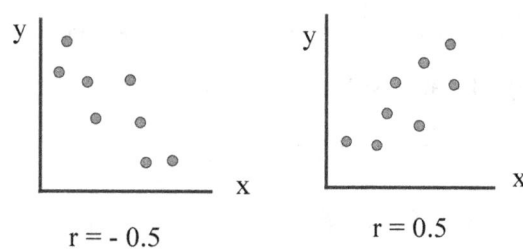

r = − 0.5

moderate
negative linear
relationship

r = 0.5

moderate
positive linear
relationship

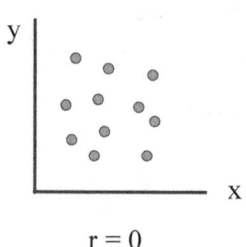

r = 0

no linear
relationship

Line of Best Fit (The Linear Regression)

(1). passing through the mean point $(\bar{x}, \bar{y})$.

(2). the difference between the model values and the real values is the least.

Use graphing calculator to find the equation of the Line of Best Fit : $y = ax + b$

e.g.

x_i	2	4	6	8	10
y_i	13	15	16	17	20

(1) Clear List L_1 and List L_2
[STAT] EDIT / 4: ClrList [ENTER] [2nd] [L_1] [,]
[2nd] [L_2] [ENTER]

(2) Enter data to L_1 and L_2
[STAT] EDIT / 1: Edit ... [ENTER]
Enter data x_i into List L_1 ;   Enter data y_i into List L_2 .

(3) Scatter Plot: [2nd] [STAT PLOT] 1: PLOT 1 [ENTER]

ON
Type:

[ZOOM] [9]

(4) Find the equation of the Line of Best Fit
and the Correlation Coefficient r :
[2nd] [CATALOG] Diagnostic On [ENTER]
[STAT] CALC / 4: LinReg(ax + b) [ENTER]
[2nd] [L_1] [,] [L_2] [ENTER]
LinReg $y = ax + b$
    a = 0.8   b = 11.4   r = 0.98

(5) Draw the Line of Best Fit:
[Y =] [VARS] 5: Statistics ... [ENTER] EQ /
1: RegEQ [ENTER] [ZOOM] [9]

(6) Predict the results by using the model:
Find the value of y when x = 10.5
[2nd] [CALC] 1: Values [ENTER]
X = 10.5 [ENTER]   Y = 19.8
(Adjust Window Dimensions for Extrapolation)

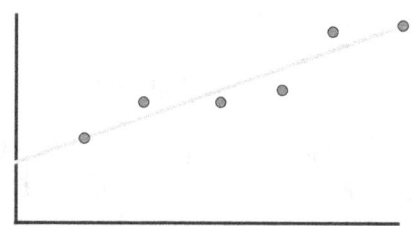

Other Regressions:

5: QuadReg: $\quad y = ax^2 + bx + c$
6: CubicReg: $\quad y = ax^3 + bx^2 + cx + d$
9: LnReg: $\qquad y = a + b\ln x$
0: ExpReg: $\qquad y = a \bullet b^x$
A: PwrReg: $\qquad y = ax^b$

To find the best model for a given set of data, compare the values of r.
Correlation Coefficient $r = \pm 1$ means exactly fit.

e.g.

x_i	0	1	2	3	4	5	6
y_i	5	10	20	40	80	160	320

(1) Enter the data into L_1 and L_2
(2) Make the scatter plot for the data
(3) Test the exponential model: $y = a \bullet b^x$
[STAT] CALC / 0: ExpReg [ENTER] [2nd] [L_1]
[,] [2nd] [L_2] [ENTER]
ExpReg $y = a \bullet b^x \quad a = 5 \quad b = 2 \quad r = 1$
$\qquad y = 5 \bullet 2^x$
(4) Draw the graph of the model:
[Y =] [VARS] 5: Statistics ... [ENTER] EQ /
1: RegEQ [ENTER] [ZOOM] [9]

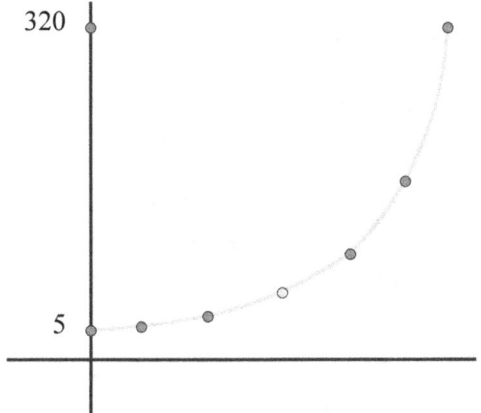

PART 4. Graphing Calculator

VIII. GRAPHING CALCULATOR

1. Tips

Clear the Memory:
[2nd] [MEM] 7: Reset ... [ENTER] 1: All Ram ...[ENTER]
  2: Reset [ENTER]
                Ram Cleared
[2nd] [MEM] 7: Reset ... [ENTER] 2: Defaults ...[ENTER]
  2: Reset [ENTER]
                Defaults Set

Return to Home Screen:
[2nd] [QUIT]

2. Graph Functions

e.g.　Graph $y = -x^2 + 4$
　　[y =] [(-)] [X, T, θ, n] [x^2] [+] [4] [GRAPH]

e.g.　Graph $y = |x - 4|$
　　[y =] [MATH] NUM / 1: abs [ENTER] [X, T, θ, n]
　　　[-] [4] [ )] [GRAPH]

Zoom Menu:
To have a better view of the graph:
[ZOOM] 6: ZStandard　$(-10 < x < 10 ; -10 < y < 10)$
[ZOOM] 4: ZDecimal　(to see friendly windows)
[ZOOM] 0: ZoomFit　(to see complete graphs)
[ZOOM] 2: Zoom In　(to see details around cursor)

[ZOOM] 7: ZTrig　　(for Trigonometry ; $X_{scl} = \dfrac{\pi}{2}$)

[ZOOM] 9: ZoomStat　(for Statistics)
e.g.　Graph $y = \sin 2x$
　　[y =] [sin] [2] [X, T, θ, n] [ZOOM] [7]

Change Window Dimensions:
e.g.　Graph $y = -3x^2 + 12x + 5$
[y =] [(-)] [3] [X, T, θ, n] [x^2] [+] [1] [2]
[X, T, θ, n] [+] [5] [ZOOM] [6]
　　To see the complete graph:
　　　[WINDOW] $Y_{max} = 20$ [ENTER] [GRAPH]

Trace:
To see the y values vary with x values:
　　[TRACE]
To find the value of y at a specific value of x:
　　[2nd] [CALC] 1: value [ENTER]

3. Table of a Function

e.g.   Display the table of $y = x^2$
    [y =] [X, T, θ, n] [x^2]
    [2nd] [TABLE]
  To change the x increment:
    [2nd] [TBLSET] ΔTbl

4. Calculation

Solve Equations:

e.g.   Solve $x^2 - 9 = 0$
 (1) graph $y = x^2 - 9$
 (2) [2nd] [CALC] 2: zero [ENTER]
 (3) move the cursor to set the Left Bound [ENTER]
   and the Right Bound [ENTER] of the x - intercept,
   then Guess [ENTER]
     Zero x = - 3   y = 0
 (4) repeat (3) to find the other zero
     Zero x = 3   y = 0

Solve the System of Equations:

e.g. solve system $xy = 8$
               $y = x + 2$

 (1) rewrite the first Eq. as $y = \dfrac{8}{x}$

 (2) graph those two functions
 (3) [2nd] [CALC] 5: intersect [ENTER]
 (4) $Y_1 = 8 / X$
   First Curve?
   move the cursor to the intersection [ENTER]
   $Y_2 = X + 2$
   Second Curve ? [ENTER]
   [GUESS] [ENTER]
   Intersection X = 2   Y = 4
 (5) repeat (4) to find the other intersection:
     X = -4   Y = -2

Maximum and Minimum:

e.g.   Find the maximum or minimum of the function
      $y = x^2 - 6x + 3$
 (1) graph $y = x^2 - 6x + 3$
 (2) [2nd] [CALC] 3: minimum [ENTER]
 (3) move the cursor to set the Left Bound [ENTER]
   and the Right Bound [ENTER] of the minimum,
   then Guess [ENTER]
     Minimum x = 3   y = -6
 when x = 3 the function has a minimum of -6 .

Sequence and Series:

 e.g. $a_n = n^2$, Find the first 10 terms and the sum.
Step 1: Enter the sequence and store it in L_1
[2nd] [LIST] OPS / 5: seq [ENTER] [X, T, θ, n]
[x^2] [,] [X, T, θ, n] [,] [1] [,] [10] [ )]
[STO >] [2nd] [L_1] [ENTER]
   Display: seq(x^2 , x , 1 , 10) → L_1
Step 2: Find the sum --- the series.
[2nd] [LIST] MATH / 5: sum [ENTER] [2nd]
[L_1] [( )] [ENTER]
   Display: sum (L_1)   385

Algebra 2/Trigonometry Reference Sheet

Area of a Triangle

$$K = \frac{1}{2} ab \sin C$$

Functions of the Sum of Two Angles

$\sin (A + B) = \sin A \cos B + \cos A \sin B$

$\cos (A + B) = \cos A \cos B - \sin A \sin B$

$$\tan (A + B) = \frac{\tan A + \tan B}{1 - \tan A \tan B}$$

Functions of the Difference of Two Angles

$\sin (A - B) = \sin A \cos B - \cos A \sin B$

$\cos (A - B) = \cos A \cos B + \sin A \sin B$

$$\tan (A - B) = \frac{\tan A - \tan B}{1 + \tan A \tan B}$$

Law of Sines

$$\frac{a}{\sin A} = \frac{b}{\sin B} = \frac{c}{\sin C}$$

Sum of a Finite Arithmetic Series

$$S_n = \frac{n(a_1 + a_n)}{2}$$

Binomial Theorem

$$(a + b)^n = {}_nC_0 a^n b^0 + {}_nC_1 a^{n-1} b^1 + {}_nC_2 a^{n-2} b^2 + \ldots + {}_nC_n a^0 b^n$$

$$(a + b)^n = \sum_{r=0}^{n} {}_nC_r a^{n-r} b^r$$

Law of Cosines

$$a^2 = b^2 + c^2 - 2bc \cos A$$

Functions of the Double Angle

$\sin 2A = 2 \sin A \cos A$

$\cos 2A = \cos^2 A - \sin^2 A$

$\cos 2A = 2 \cos^2 A - 1$

$\cos 2A = 1 - 2 \sin^2 A$

$$\tan 2A = \frac{2 \tan A}{1 - \tan^2 A}$$

Functions of the Half Angle

$$\sin \frac{1}{2} A = \pm \sqrt{\frac{1 - \cos A}{2}}$$

$$\cos \frac{1}{2} A = \pm \sqrt{\frac{1 + \cos A}{2}}$$

$$\tan \frac{1}{2} A = \pm \sqrt{\frac{1 - \cos A}{1 + \cos A}}$$

Sum of a Finite Geometric Series

$$S_n = \frac{a_1(1 - r^n)}{1 - r}$$

Normal Curve
Standard Deviation

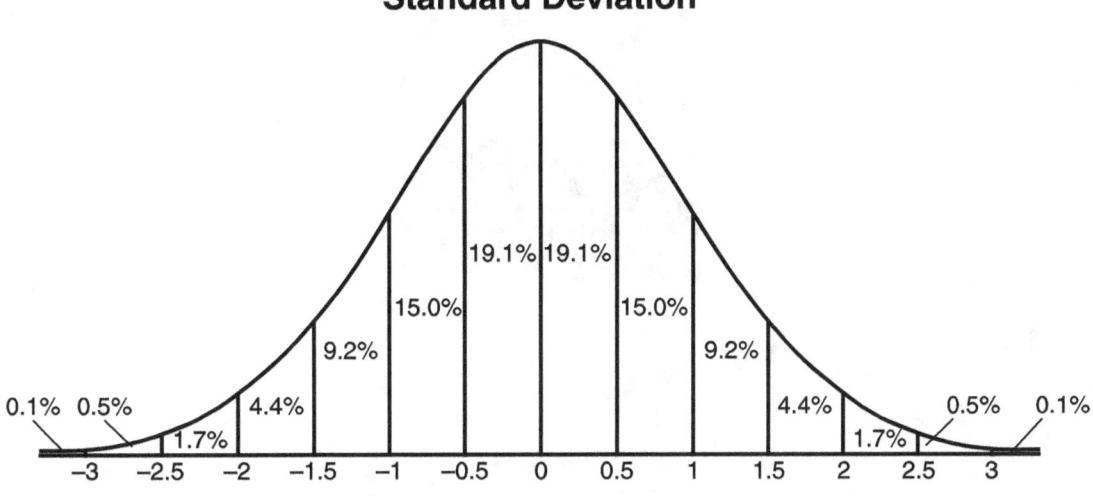

Also Available

Student's Choice
Regents Review Integrated Algebra   ISBN: 9781453880982

Student's Choice
Regents Review Geometry   ISBN: 9781453709993

Student's Choice
Regents Review Algebra 2/Trigonometry   ISBN: 9781460983874

Coming Soon

The Calculus Handbook: Concepts and Skills